# The search for gravity waves

# The search for gravity waves

P. C. W. DAVIES

*Professor of Theoretical Physics, University of Newcastle-upon-Tyne*

CAMBRIDGE UNIVERSITY PRESS

*Cambridge*

*London   New York   New Rochelle*

*Melbourne   Sydney*

Published by the Press Syndicate of the University of Cambridge
The Pitt Building, Trumpington Street, Cambridge CB2 1RP
32 East 57th Street, New York, NY 10022, USA
296 Beaconsfield Parade, Middle Park, Melbourne 3206, Australia

First published 1980

Printed in the United States of America
Photoset in Malta by Interprint Limited
Printed and bound by Vail-Ballou Press, Inc., Binghamton, New York

*British Library cataloguing in publication data*
Davies, Paul Charles William
The search for gravity waves.
1. Gravity waves
I. Title
531'.14   QA927   79-42616
ISBN 0 521 23197 3

# CONTENTS

# PREFACE

The search for gravity waves is an exciting and somewhat bizarre episode in the development of science and technology in the past twenty years. Few physicists seriously doubt that waves in the gravitational field, analogous to waves in the electromagnetic field, really exist. However, calculations indicate that although the emission of gravitational radiation by distant astronomical objects could have a vastly greater impact on their evolution and structure than their electromagnetic counterpart, only something like $10^{-76}$ of the energy released is likely to register its presence in laboratory detectors on Earth. This is because, notwithstanding the colossal quantities of energy sure to be contained in the larger gravity wave outbursts in space, the interaction of gravitational radiation with matter is minute.

The extreme weakness of the sought-for effects demands a technology of dazzling capabilities. The detection of just one quantum of vibration in a tonne of metal is being planned. Movements of only $10^{-21}$ m in a highly refrigerated metre-long bar must be measured. Great strides towards achieving these extraordinary accuracies have been made, and the development of gravity wave detectors is proceeding apace. A pivotal event in this programme occurred in the early 1970s when Professor Joseph Weber of the University of Maryland claimed to have discovered gravity waves in the first detector. Although subsequent work has not confirmed these results, the establishment of a new branch of astronomy, using gravity wave detectors as 'gravity telescopes', is on the horizon. With such a facility we could 'see' into the dense hearts of quasars and neutron stars, probe to the very edges of black holes and maybe eventually listen to the rumble of the primordial big bang itself.

An added impetus to this fascinating development came with the discovery of the so-called binary pulsar in 1974. This object displays unmistakable signs of emitting gravity waves.

It is therefore timely to give an account for non-specialists of the subject of gravity waves and their detection. No advanced knowledge of physics or astronomy is needed to understand this book. The subjects of gravity and Einstein's theory of relativity are explained from basics; mathematics is kept to a minimum, employing only elementary high school algebra and calculus. In many cases I have used words in equations rather than resort to a proliferation of formal symbols. The level of exposition corresponds roughly to that of *Scientific American* or *New Scientist*.

The treatment is not intended to constitute a textbook, but rather a survey of a scientific adventure story that promises to bring a rich harvest of rewards in the coming decades.

I am grateful to N. D. Birrell, S. A. Huggett, M. J. Rees, D. C. Robinson and W. G. Unruh for helpful information and comments. I should also like to thank S. Mitton, G. A. Papini and J. Weber for supplying photographs.†

**Note on units and nomenclature**. In this book I have used the internationally accepted system of units (SI units), except occasionally in the discussion of astronomical distances, where the parsec $(=3.09 \times 10^{16} \text{m} = 3.26$ light years), abbreviated pc, or astronomical unit (average Earth-orbit radius $= 1.50 \times 10^{11}$ m), abbreviated AU, are more appropriate. Where 'billion' is used, the USA value of $10^9$ (one thousand million) is intended.

Many of the numerical estimates and some of the formulae quoted are only accurate to within an order of magnitude, in which case the equality sign is replaced by $\sim$. In other cases it has proved convenient for exposition to round off numerical factors, and in these cases the approximate equality sign $\approx$ is used.

*Newcastle-upon-Tyne* P. C. W. Davies

---

†Since this book went to press, Professor Papini's experiment has been discontinued.

# 1    Electromagnetic waves

Most people are familiar with electric and magnetic forces from daily life or elementary laboratory experiments. What can be hard to understand is how these forces can be translated into wave motion which can leave the laboratory and travel off into space as an apparently independent entity.

Sound waves and water waves do not seem so remarkable because we can detect a tangible medium which vibrates when the wave passes. An electromagnetic wave, on the other hand, is not a disturbance in any substance, and can travel unimpeded through a perfect vacuum. Heat and light radiation from the Sun – perhaps the two most familiar electromagnetic waves – reach the Earth across 150 million kilometres of empty space.

What are electromagnetic waves? In the following sections it will be explained how electric and magnetic forces can operate on each other to produce wavelike disturbances in free space. Some of the properties of these disturbances will also be described for later comparison with gravity waves.

## 1.1    Forces and fields

Electricity manifests itself as a force which acts on electrically charged bodies. The simplest examples of such a force are the so-called electrostatic effects, such as occur when combing one's hair or when a stroked rubber balloon adheres to the ceiling.

Electric charges come in two types, called positive and negative. Like charges repel each other, but unlike charges attract. Both forces diminish rapidly with distance between the charges. The existence of two varieties of electricity, and hence both attractive and repulsive forces, is in contrast to gravity, which always attracts.

It is now known that all electricity is fixed to subatomic particles in definite multiples of a fundamental quantity – an 'atom' of electricity. Not all types of subatomic particles carry electric charge, but of those that do, the lightest is the electron and this carries one unit of charge. All normal atoms contain electrons, and often they can be fairly easily detached. In some substances, such as metals, free electrons are prolific. It is usually the rearrangement of electrons that causes a macroscopic body to become electrically charged.

A normal atom has equal quantities of both positive and negative electricity, so it is electrically neutral. The number of 'units' of electricity carried by all the atom's electrons, which by convention is taken to be negative, is exactly balanced by an equal and opposite positive charge located on the atomic nucleus. If a body contains a surfeit of electrons it will become negatively charged, while a deficit of electrons implies some unbalanced positive charges on the nuclei, and hence the body will be overall positively charged.

Electrons are fairly mobile particles, and travel easily through most metals. Under the action of electric forces, electrons repel each other, so there is a natural tendency for a local aggregation of them to disperse themselves. For example, if a crowd of electrons is located at the end of a metal wire, the electrons will flow along the wire to escape their mutual repulsion. In this way electric currents occur.

In the early nineteenth century it was discovered that electric currents produce magnetic forces and that, conversely, if a magnet is waved about near a wire, then a current can be induced to flow (see Fig. 1.1). This interplay between electric and magnetic forces is

Fig. 1.1. Electromagnetic induction. If the magnet is suddenly withdrawn from the interior of the coil, the change in magnetic field threading the wire causes an electric field which drives a current round the coil.

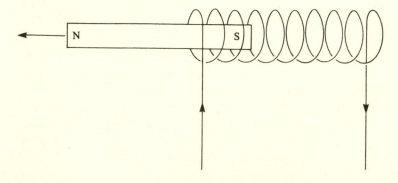

central to an understanding of electromagnetic waves. Today it is believed that all magnetism is caused by electric currents; i.e., there are no 'atoms of magneticity' that play the role for magnetism that electrons play for electricity. In the case of the Earth's magnetism, for example, there are electric currents deep in the planet's interior. The magnetism of an ordinary bar magnet is due to circulating currents at the molecular level.

One convenient way of describing the forces which act between electric charges and currents is in terms of the *field* concept, first introduced by Michael Faraday. Rather than say that two charges attract (or repel) across empty space, one says that every electric charge produces an electric field around itself, the strength of which diminishes with distance. The force that is experienced by a nearby charge is then attributed to the interaction between the latter charge and the field. Of course, this charge is also the source of its own field, which will react back on the former charge as well. The strength of the force is proportional to the strength of the field at that point. These ideas are depicted in Fig. 1.2.

The advantage of the field concept is that the interaction between separated charges is reduced from a non-local action-at-a-distance, to a local charge–field interaction. Magnetic fields may be introduced similarly.

It is possible to give meaning to the *shape* of the field as a means of representing the pattern of force around a particular configuration of charges of magnets. Fig. 1.3 shows the force field around a single point electric charge. The radially symmetric lines

Fig. 1.2. Electric field. The fixed positive charge is surrounded by an invisible electric field. The test charges sense the field in their vicinity and are driven towards (−) or away from (+) the central charge.

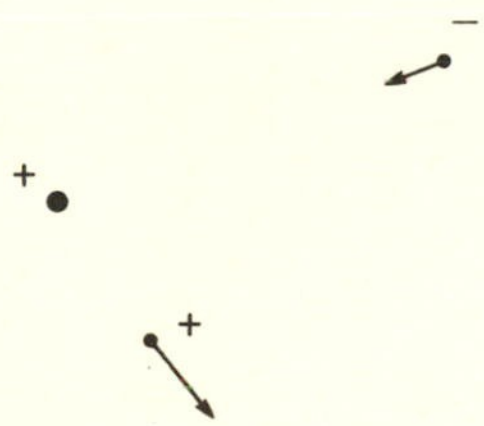

are so-called 'lines of force' and they represent the direction of force acting on a positive charge placed at that point. When the nature of the electromagnetic field is being probed using charged particles, we have in mind a so-called test charge which, although it responds to the presence of the field, does not itself react back electrically on the field. Thus the test charge does not disturb the system being investigated. In practice this situation can be well approximated by using a test particle with a very small charge. Similarly, when we come on to the subject of the gravitational field, test masses, whose reaction on the gravitational field may be neglected, will also be discussed. As shown, the effect is a repulsion directly away from the central charge, as indicated by the outwardly directed arrows. If the central charge were negative the arrows would point inwards, indicating a radial attraction.

A measure of the force is provided by the relative density of lines. Near the centre, where the force is strong due to the close proximity of the charge, the lines are crowded, but they fan out with distance, indicating a weakening of the field and its corresponding force on a test charge. As lines of force cannot end, except on other charges, the total number is constant. If one considers concentric spheres about the central charge, then each sphere is threaded by the same number of lines. As the surface area of the spheres increases like (radius)$^2$, the density of lines diminishes with distance like 1/(radius)$^2$, or an inverse square law.

Fig. 1.3. Lines of force. The radial pattern diverging from the positive charge maps the shape of the electric field. As the lines fan out, their density falls off inversely as the square of distance from the charge (in three dimensions), indicating an inverse square law of force.

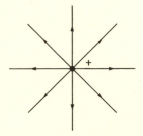

Fig. 1.4 shows the field lines around a more complicated charge system, and the analogous lines of magnetic force around a bar magnet. Magnetic field lines represent the force acting on a point test north pole.

To the physicists of the nineteenth century, electric and magnetic fields assumed an almost tangible status, and were envisaged as a sort of invisible fluid medium. The mysterious, ephemeral medium was given the name aether, and was supposed to fill all of space. Electric and magnetic fields were then identified with stresses in the aether medium.

It is clear that the aether cannot be a medium of a familiar sort, for electrically neutral material bodies may pass through it without encountering any resistance. For example, the Earth orbits the Sun in apparently drag-free conditions.

## 1.2    Electromagnetism

As a simple conceptual aid, electric and magnetic fields are invaluable, but at first sight they appear to be redundant physically. Nothing as yet separates the behaviour of the field from the electric and magnetic sources to which it is tied. We could either describe the forces as due to electric charges and currents acting at a distance, or regard them as sources of fields and look to the fields for an explanation of electric and magnetic force. It is merely a matter of linguistic convenience.

All this changed with the revolutionary work of James Clerk

Fig. 1.4. Dipole fields. (*a*) The conjunction of equal and opposite (+, −) charges is called a dipole and produces a static field with a complicated structure. The test charge *A* moves obliquely to both + and −. (*b*) A magnetic dipole (bar magnet) formed from the conjunction of two magnetic poles (N, S; north, south) has a similar magnetic field shape.

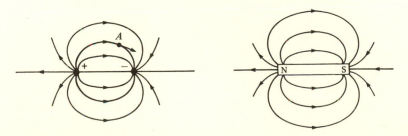

Maxwell in the early 1860s. Maxwell's achievement provides a beautiful example of how mathematical symmetry and elegance can be employed to improve our understanding of nature.

The fact that magnetic fields are produced by, and can act on, electric currents indicates a deep connection between electricity and magnetism. There are, however, two ways in which the relationship between them seems lopsided. The first is the absence of magnetic charge – magnetic fields seem only to be produced by electric currents. This asymmetry has long been a puzzle to physicists, and some believe that magnetic charges do exist on hitherto un-discovered subatomic particles, but there is no experimental evidence to support this conjecture. The second lopsidedness between electricity and magnetism is that, while a changing magnetic field will induce an electric field that can make an electric current flow (see Fig. 1.1), the reciprocal effect was not known in the mid nineteenth century. Maxwell puzzled over this because the equations which connect the strength of a field to the behaviour of its sources are inconsistent unless a changing *electric* field can induce a *magnetic* field, as well as vice versa.

To achieve mathematical consistency, Maxwell introduced a new term into the field equations, representing the missing effect. Thus, one of the asymmetries in the theory was removed. More importantly, Maxwell's bold step transformed the nature of the fields. If a changing electric field can induce a magnetic field, then as this latter field builds up, it will in turn induce an electric field. But the build-up of the new electric field goes on to produce its own changing magnetic field, and so on. The exciting possibility arises that changing electric and magnetic fields can sustain each other in a sort of perpetual motion. Moreover, because each field acts as a source of the other, the fields can even exist and move in regions of space where there are no electric charges or currents to act as sources. The fields therefore acquire an independence that was quite unsuspected before Maxwell. No longer tied to charges and currents, the fields are freed to assume a separate mechanical existence. They have been elevated from the status of a linguistic convenience, to that of a real, independent, physical system.

Because the self-sustaining electric and magnetic disturbances always require both electric and magnetic fields together – each to feed off the other – we are really dealing with a single, unified, *electromagnetic* field, of which electric and magnetic fields individually are merely components. Maxwell investigated the behaviour of these self-sustaining electromagnetic motions, and soon discovered that one particularly simple solution to his equations exists. The pattern of field motion can assume a very familiar form – that of a wave. The equations indicated that the speed of the wave depends on the electric and magnetic properties of the medium in which it propagates. In free space, the speed works out at around $3 \times 10^8 \mathrm{ms}^{-1}$, which is the speed of light. Maxwell concluded that light is an electromagnetic wave, and thereby achieved a brilliant synthesis of the science of optics with that of electromagnetic theory.

## 1.3    Electromagnetic waves

What is an electromagnetic wave? In the nineteenth century it was fashionable to envisage the wave disturbance as a vibration of the mysterious aether, rather as a sound wave is a vibration of the air. As we shall see, this picture is both unnecessary and, if taken too literally, incorrect.

An electromagnetic wave is primarily an undulation of electric and magnetic force. If we regard it as a moving, changing field, we can map how the field changes by stationing electric charges in its path to measure the local field strength. Because all *periodic* undulations of a linear system can, by the theorem of Jean Fourier, be built up from a superposition of pure *harmonic*, or sinusoidal, waves, we need only consider a wave motion of the form $\sin(\omega t + \phi)$ where $\omega$ is the angular frequency of the oscillations, $t$ is the time and $\phi$ is a constant phase angle.

Fig. 1.5 shows the effect on a positive electric test charge caused by the passage of an electromagnetic wave travelling perpendicular to the page. As the wave passes, the magnitude of the electric field **E** rises and falls in the sinusoidal fashion shown. This field exerts a force on the test charge and wiggles it up and down periodically. At time zero, there is no electric field and no force. Then the field starts to build up strength, peaking at $\frac{1}{4}$ cycle, driving the charge upwards with its greatest force. The field then starts to decline and the upward force falls away until at $\frac{1}{2}$ cycle it vanishes. Thereafter the direction of the field and the force is reversed as we move down into the 'trough' of the wave. The particle is forced downwards. Once again the field reaches maximum intensity at $\frac{3}{4}$ cycle, and then declines, until after one cycle it has dwindled to zero again and is ready to repeat the next cycle.

It is worth noting that the *motion* of the test charge is $\frac{1}{2}$ cycle out of phase with the driving force; i.e., when the force is directed upwards, the particle is moving downwards, but decelerating. It reaches the bottom of its trajectory when the upward force is a maximum. The arrows in Fig. 1.5 depict the *force*, not the motion.

The words 'upwards' and 'downwards' have been used purely schematically, for the electromagnetic wave need have no relation to up or down. Indeed, the undulating electric field can point in any direction perpendicular to the direction of propagation of the wave (see Fig. 1.6). Waves of this variety are called *transverse*. This can be expressed in terms of vectors (directed quantities denoted by arrows). In Fig. 1.6 the vector **k** marks the propagation direction of the wave, while the unit vector **e** represents the electric force direction. As the wave is transverse, **k** and **e** are perpendicular.

Fig. 1.5. Electromagnetic wave. The electric field undulates sinusoidally as shown below. The positive test charge (above) experiences the changing forces indicated. The absence of an arrow implies instantaneously zero force.

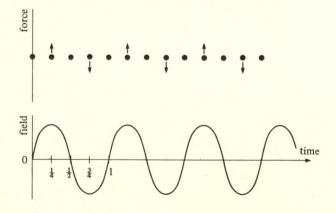

The direction of **e** is known as the *polarization* vector. Light from an ordinary source will contain electromagnetic waves of many different polarization directions all mixed together, but if it is passed through a polarizer, only waves vibrating in one particular direction will be passed, the others being filtered out.

A mathematical analysis shows that the magnetic field **B** oscillates with the same frequency as the electric field, but in a perpendicular direction (see Fig. 1.7). It is interesting to consider the effect of both the electric and magnetic forces on the test charge. When the charge starts to move under the action of the electric field, it constitutes a tiny electric current, which has *magnetic* action (this is the principle of the electric motor, where a current-carrying coil is forced to rotate in the field of a magnet). The magnetic force acting on a current is perpendicular to the current and to the applied magnetic field, so here it is directed along the line of propagation (i.e., along **k**). The main effect of the wave is to wiggle the test charge perpendicular to **k**, but as a small secondary effect it will also drive the charge along somewhat. If the charged particle is subject to some damping forces, it will oscillate slightly out of phase (i.e., it will lag behind the driving field), and the average effect

Fig. 1.6. Transverse wave. The wave travels forward in the direction of vector **k**, but the electric (and magnetic) fields are transverse to this. The electric field undulates along the direction of **e**, perpendicular to **k**. The test charge (blob) is driven back and forth along direction **e** by the oscillating electric field.

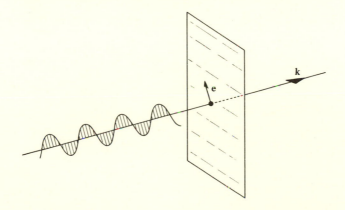

of the action along **k** will be the exertion of a pressure. In other words, the charged particle *recoils*, and we see that the electromagnetic wave must carry momentum, some of which is imparted to the charge.

That electromagnetic waves carry momentum and can exert a force when they strike matter is beautifully illustrated by the tails of comets, which consist of gas that is literally blown out of the cometary head by the pressure of sunlight. Ambitious proposals have been made to 'sail' a spacecraft to distant planets using this light pressure as a driving force.

Clearly, if the charged particle is set in motion it acquires energy, which tells us that the wave carries energy as well as momentum. Maxwell's theory shows that the energy density is simply, in Gaussian units,

$$\frac{1}{8\pi}(\mathbf{E}^2 + \mathbf{B}^2)$$

where **E** and **B** are the electric and magnetic field strengths respectively, while the momentum is

$$\frac{c}{4\pi}(\mathbf{E} \times \mathbf{B}),$$

Fig. 1.7. The electric (**E**) and magnetic (**B**) fields oscillate in phase, perpendicular to each other and to the propagation direction **k**. When a test charge moves along **E**, it makes a current which the magnetic field **B** forces in the direction **k**.

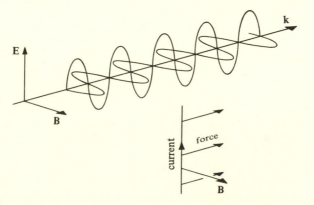

the vector product of the two, multiplied by $c$, the speed of the wave.

It was mentioned above that any periodic waveform can be built up from sine waves of various frequencies. This is only true if all the constituent waves have the same direction of polarization; we cannot build a wave which vibrates, say, east–west, from waves which vibrate north–south. To take into account this additional requirement of polarization it is necessary to build up a general wave by using sinusoidal waves belonging to two independent sets, each with polarization perpendicular to the other. Any intermediate polarization direction can then be built up by vector addition (see Fig. 1.8).

As remarked, it is possible to create waves that are polarized in one particular direction. Another possibility is to combine together two perpendicularly polarized waves in a different way, as shown in Fig. 1.9. Here the waves are set $\frac{1}{4}$ cycle ($\pi/2$ radians) out of phase, so that when one wave reaches peak field strength the other is zero, and vice versa. This means that if the waves are of equal strength, and the time dependence of one wave is described by $\mathbf{e}_1 \sin \omega t$, the other is described by $\mathbf{e}_2 \sin(\omega t + \pi/2) = \mathbf{e}_2 \cos \omega t$. As $\mathbf{e}_1$ and $\mathbf{e}_2$ are perpendicular, the *strength* of the resultant superposed wave is, from the rule of vector addition,

$$\sqrt{(\mathbf{e}_1{}^2 \sin^2 \omega t + \mathbf{e}_2{}^2 \cos^2 \omega t)} = 1$$

Fig. 1.8. Superposition of polarized waves. Adding two unequal strength waves with perpendicular polarizations produces a wave with intermediate polarization vector **e**. The magnetic fields are not shown.

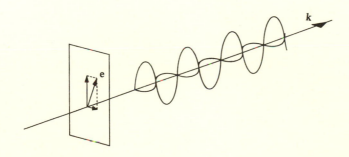

because $e_1{}^2 = e_2{}^2 = 1$ (unit vectors). The net field strength is therefore constant – it does not oscillate. However, the *direction* $\theta$ of the resultant vector **e** changes (see Fig. 1.9)

$$\tan \theta = \frac{\sin \omega t}{\cos \omega t} = \tan \omega t$$

or

$$\theta = \omega t.$$

The angle $\theta$ increases uniformly with speed $\omega$, i.e., the polarization vector **e** rotates anticlockwise with uniform angular frequency $\omega$. This arrangement is called circular polarization, and if a circularly polarized wave hits some test charges it will tend to twist them. The torque exerted shows that electromagnetic waves carry angular momentum, or spin, as well as linear momentum and energy.

Fig. 1.9. Circular polarization. If equal strength waves are superposed $\pi/2$ out of phase, the direction of the resultant electric field rotates, once with each cycle of the wave, but the net strength of the field is constant – it does not oscillate. In the quantum theory (below) this can be envisaged as a photon, spinning as it travels.

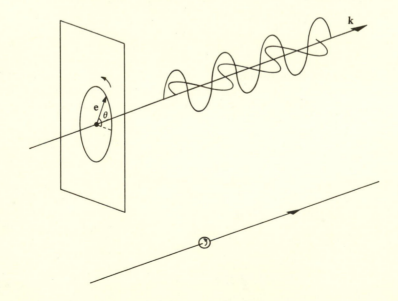

In the quantum theory (see section 5.4) light is regarded as composed of 'photons, which in some circumstances can be regarded as particles. It is sometimes helpful to envisage these particles as carrying energy and momentum, and also to be spinning around their direction of motion (see Fig. 1.9) as they travel. Quantum theory gives the value of these quantities in terms of Planck's constant $h$ and frequency $v$:

$$\text{energy} = hv$$

$$\text{momentum} = \frac{hv}{c}$$

$$\text{spin} = \frac{h}{2\pi}.$$

Physicists usually measure atomic spin in units of $h/2\pi$, denoted $\hbar$, so in this system the photon has spin 1. Later these considerations will be extended to gravitational waves.

## 1.4 Sources of electromagnetic waves

Given that electromagnetic fields can exist and propagate as waves through empty space, devoid of electric charges or currents, the question arises as to how this self-sustaining field motion is generated in the first place. Where do the waves come from? Once started, they can undulate of their own accord, like waves on a pond that persist long after a stone has been thrown into the centre. But what plays the role of the stone?

There is an interesting question about whether the Universe was created with electromagnetic ripples built in at the outset, but leaving aside issues of cosmology the obvious way to generate undulating fields is to use electric charges and currents. The fact that light is an electromagnetic wave shows that there must be electricity in all luminous matter, and we now know that all atoms contain electric particles that can emit and absorb photons when they are disturbed.

Several years after Maxwell's work, Heinrich Hertz succeeded in producing electromagnetic radiation in the laboratory. Unlike light, the wavelength of which is around $10^{-7}$ m, Hertz's waves were many metres long, a region of the electromagnetic spectrum that we now call radio. Today, it is recognized that gamma- and X-rays, ultra-violet and infra-red heat are also electromagnetic waves, differing from each other, and from light and radio waves, only in wavelength and frequency. They all travel at the speed of light ($c$) in a vacuum. Fig. 1.10 shows the entire electromagnetic spectrum.

It is easy to understand qualitatively how electromagnetic waves can be produced using electric charges. A point charge at rest is surrounded by a radial electric field (see Fig. 1.3). Suppose the charge is moved suddenly a little to one side. The surrounding field will have to adjust itself to the new location of the charge; the lines of force will have to radiate from a slightly displaced position. However, we know from Maxwell's theory that a disturbance in the field travels at a fixed speed through empty space – the speed of light, $c$. Consequently, the distant regions of the field will not know of the relocation of the charge until a time $r/c$ later, where $r$ is the radial distance from the charge. Thus, the nearby regions of the field respond rapidly, and the field lines can be rearranged to centre on the new position of the charge, but the distant field will be as yet unaffected. It follows that where the two regions join there will be a sudden kink in the field (see Fig. 1.11). This kink, or wavelike disturbance, travels outwards from the charge at the speed of light. Note that, at the kink, the field changes direction by 90°. Whereas the undisturbed field acts radially *outwards* on a test charge, the wavelike disturbance is *transverse*, as expected for a radiation field.

Recall that the density of field lines is a measure of the strength of the field at that point. At the kink, the distorted lines crowd into a thin shell. (The shape of the lines in the shell reflects the detailed motion of the charge during the brief period of its displacement

Fig. 1.10. Electromagnetic spectrum.

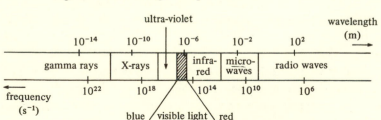

from the first location to the second.) The field is therefore enhanced within the shell region over the surrounding regions. The
shell volume grows like $r^2$, but the length of each kink grows like $r$,
because the mismatch of the two radial patterns is proportionally
greater at larger distance $r$ from the charge. The net effect is that
the kink field falls off like $1/r$ in strength, rather than $1/r^2$ as for the
undisturbed radial field. This slow rate of decline is characteristic of
electromagnetic radiation, and tells us that the disturbance can
propagate to great distance with appreciable strength.

In fact, recalling that the energy density of the field contains the
factor $\mathbf{E}^2$, we see that the field energy density falls off as $1/r^2$, which
means that the total energy crossing a spherical surface centred on
the charge is constant, independent of the size of the sphere. This
means that the radiation disturbance conserves energy as it travels
outwards.

Fig. 1.11. Electric kink. The sudden displacement of the
(positive) electric charge sets up a kink in the field which propagates outwards at the speed of light. Inside the kink the lines
of force radiate from the new location, outside they radiate
from the old location. In the kink itself the lines are dense
(strong field) and perpendicular to the radial lines (transverse
field). The kink is greatest near the perpendicular to the direction of the line of motion of the charge and least near this direction ($\sin^2 \theta$ angular dependence of energy).

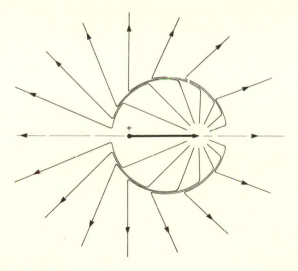

Another feature of electromagnetic radiation can be deduced from Fig. 1.11. The size of the kink is obviously somewhat less near the direction of the charge's displacement than perpendicular to it. Elementary geometry indicates a factor $\sin \theta$, where $\theta$ is the angle of the direction of interest from the line of motion of the charge. The radiated energy therefore looks like the shape shown in Fig. 1.12.

If the charge were moved back again suddenly, another kink would occur. Continued agitation produces continuous disturbance in the field, and if the motion is smooth and periodic, the radiated kinks will take on the features of an undulation, or wave. A sinusoidal vibration of the charge will generate a pure harmonic wave of the same frequency. To bring about such a vibration of the charge, it must be confined and moved by a suitable force, such as another electric field. In a radio transmitter, electrons are driven along wires by an applied electromotive force. When one end of the system is charged negatively with these electrons, the other end is charged positively and vice versa. So when electrons wiggle about in ordinary matter the matter becomes *polarized*. The situation depicted in Fig. 1.11 can be realized in practice by placing a metal rod (aerial) along the axis, and driving the electrons back and forth with a voltage. The resulting polarization pattern is called a dipole because there is negative charge at one end, and positive charge at the other. The radiation pattern is identical whether only the negative charge moves, or only the positive charge, or both. The situation could be idealized by a single dipole consisting of two

Fig. 1.12. Angular distribution of radiation energy round a dipole source.

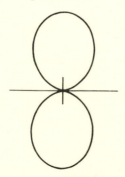

charges, one positive, one negative, fixed in a sort of dumb-bell arrangement, oscillating along the dumb-bell axis.

A more complicated radiation pattern will be produced by more elaborate arrangements of moving charges. Another idealized situation is depicted in Fig. 1.13. This is a pair of dipoles side by side and opposed. Set this system into oscillation and a more complicated pattern results, known as quadrupole radiation after the four poles (charges) shown in the figure.

Higher pole radiation may be built up similarly. If we take an arbitrary distribution of radiating charges we may regard the resulting radiation as composed of dipole, quadrupole, octupole, and so on, all superimposed. If the geometry of the source is fairly simple, the radiation will be dominated by the lower pole contributions, the admixture of complicated higher pole radiation being very weak.

It is interesting to note that the shape of the radiation pattern gives information about the internal structure of the source. Analysis of the multipole nature of gamma rays has been used to explore the organization of atomic nuclei, which emit high energy photons when their protons are rearranged. Later we shall see how multipole considerations apply to gravity waves.

Of course, the construction based on Fig. 1.11 is very heuristic. Nothing was said, for example, about magnetic fields. A proper treatment of the emission of radiation by a distribution of electric charges and currents requires a detailed analysis of Maxwell's equations. This treatment is beyond the scope of the book, but we shall end this section by giving a brief summary of the basic structure of Maxwell's equations.

Maxwell combined together a number of laws concerning electri-

Fig. 1.13. This symmetric arrangement of four charges (two dipoles) is called a quadrupole.

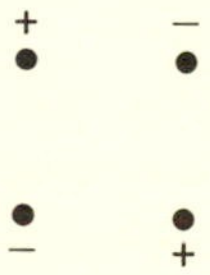

+     −
●     ●


●     ●
−     +

city and magnetism into a single system of equations. The first of
these, known as Faraday's law, describes the variation from place
to place of the electric field induced by a changing magnetic field.
In words,

spatial variation of **E** = time rate of change of **B**.

The second equation embodies André Ampère's law on the way in
which electric currents produce magnetic fields, but augmented
with Maxwell's additional term that describes how a changing
electric field also produces a magnetic field:

spatial variation of **B** = electric current
+ time rate of change of **E**.

Without the final term, the equations are inconsistent.

A third equation embodies the observation that free magnetic
poles do not exist, so that the lines of magnetic force cannot have
ends. The final equation, which is called Coulomb's law after
Charles Coulomb, relates the appearance of new lines of electric
force to the density of electric charges in a region of space – electric
force lines begin (or end) on free electric charges. These two
equations can be expressed as
number of new lines of force
emerging from a region of space = 0 (for magnetic field)
= density of electric charge (for
electric field)

The task is to find solutions for these equations which give the
strengths of the electric and magnetic fields **E** and **B** in terms of the
currents and charges. By taking the time rate of change of the
second of Maxwell's four equations, the magnetic field can be
eliminated using the first equation. What results is a new equation
connecting the second derivatives of **E** (i.e., the variation of the
space variation, and the time rate of change of the rate of change)
and the rate of change of electric current. A similar equation
obtains for **B** when **E** is eliminated.

These equations that couple the spatial and temporal variations
of the field are characteristic of wave equations, and possess
solutions that are both harmonic in space (a sinusoidal wave
pattern) and in time (periodic undulations of sinusoidal strength). If
the current is chosen to be an oscillating dipole, solutions can be

found that describe a $1/r$ type radiation field with a $\sin\theta$ (dipole) angular dependence, superimposed on top of the ordinary $1/r^2$ electric fields. These solutions show that the wavelike magnetic field falls off like $1/r$ also, and that far from the source, the wave disturbance approaches the form discussed in the previous section, with **E** and **B** perpendicular, and transverse to the propagation vector **k**.

## 1.5   The special theory of relativity

As mentioned, one of the triumphs of Maxwell's theory was that his electromagnetic waves turned out to travel with the same speed as light, from which he concluded that light is a vibration of the electromagnetic field. But the notion of speed is ambiguous. What is the speed of the Earth? The Sun? The Galaxy?

How do we measure speed? The speed of an aeroplane might be $200$ m s$^{-1}$ to the people on the ground, but it is zero to the passengers. When we say a plane is travelling at $200$ m s$^{-1}$, we really mean '$200$ m s$^{-1}$ relative to the ground' or perhaps to the air. Similarly, the speed of a car is measured relative to the road.

If a body moves uniformly there is no way that any mechanical effect, *within the body*, can be used to measure its speed. Only by observing one's environment can the speed relative to one's surroundings be deduced. If those surroundings are also moving, then yet another system is needed to gauge its motion, and so on. There is no visible system or body in the universe that we can regard as absolutely at rest, against which all speeds can be measured in an absolute way. Of course, if the motion of a body is not uniform, such as when an aeroplane hits turbulence, absolute effects occur, such as coffee spilling into passengers' laps. But if the motion is unaccelerated, only *relative* speed is meaningful mechanically.

What then, is the speed of Maxwell's waves? His equations give a unique answer: $3 \times 10^8$ m s$^{-1}$. No provision is made in the theory for who might be measuring the speed, or what their own speed might be.

In the nineteenth century, physicists supposed that the speed of light referred to the motion of the waves relative to the aether, which defined a state of absolute rest against which all other motion could be gauged. The Earth, for example, was envisaged as sweeping through the aether much as a fish swims through the sea. Unlike the fish, though, the Earth experiences no friction, nor does the aether exert any mechanical effect on bodies moving through it with uniform speed. If it did, the principle of relativity of motion, on which all of mechanics is founded, would have to be abandoned.

In the early years of this century, Albert Einstein investigated the dynamics of electric particles, paying special attention to the question of the relativity of motion. Einstein regarded the principle of relativity as sacrosanct, and wished to extend it to electromagnetic theory. He proposed that the aether concept be replaced by an extraordinary new law of nature. The unique speed $c$ which appears in Maxwell's equations is to be taken as applying to every observer, irrespective of their state of motion. What this means is that an observer will measure the speed of a light pulse to be $c$, and another observer, racing past him, will measure the speed of the *same pulse* to be $c$ also! It follows that if you chase after a light pulse in a rocket, then, however powerful are the motors, you will not gain even one metre per second on the receding pulse.

The bizarre nature of this proposal is apparent if one observer remains on the Earth, watching the light pulse recede into space, the pulse being chased by his colleague in the rocket (see Fig. 1.14). As the light recedes at the fixed velocity $c$, the Earth man will

Fig. 1.14. Speed of light is constant. The rocket chases the light pulse. The astronauts see it continuing to recede at $c$ relative to the rocket, even when on full power. From Earth however, the light pulse appears to recede at $c$ relative to *Earth*, so the rocket does appear to be making some progress in reducing the rate of recession of the pulse relative to the rocket.

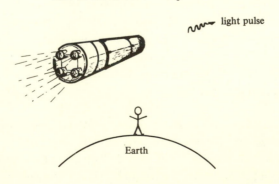

light pulse

Earth

obviously see the rocket reducing the speed of recession of the pulse relative to the rocket as the rocket accelerates after it. So although, when seen from Earth, the light is receding from the rocket at less than $c$, from the rocket it still recedes at $c$. To reconcile these viewpoints it is necessary to suppose that the gap between rocket and pulse appears, from the rocket, to be shrunken by a factor $(1 - v^2/c^2)^{\frac{1}{2}}$, where $v$ is the rocket velocity. The idea that the distance between two points can be different for different observers is quite without precedent in physical science. Space was formally regarded as a sort of rigid arena against which bodies can be located, but itself an essentially passive entity, not subject to mutation or distortion.

If travelling fast makes the miles shrink, it is clear that rocket people arrive at their destination somewhat sooner than otherwise expected. It follows that with the unshrunken perspective, from Earth, the observed journey of the rocket must appear to take longer than the *same* journey appears to take to observers in the rocket. This implies that the time scale is different in the two systems, and the rocket clock runs slow relative to the Earth clock, again by the factor $(1 - v^2/c^2)^{\frac{1}{2}}$. Thus both space and time are 'elastic', and can be shrunk or stretched depending on one's state of motion.

Before Einstein's work, great importance was attached to attempts to measure the speed of the Earth through the aether using light rays. In one famous experiment, performed by Albert Michelson and Edward Morley in 1887, the speed came out to be zero, implying that the Earth was at rest! With the benefit of Einstein's new theory of relativity, published in 1905, this result is expected, for light rays will always travel with velocity $c$ relative to the experimenter, whatever his motion.

Since then, Einstein's ideas have been tested experimentally in many ways, and effects such as time stretching can be measured directly. In a recent experiment, subatomic particles called muons were created and stored in a magnetic ring, moving at near-luminal velocity. Their lifetime against radioactive decay was observed to be prolonged by a factor of 29.33. The theory of relativity pervades much of modern science, and is now part of mainstream physics.

One important consequence of the theory is obvious: because

one cannot gain at all on a light pulse by accelerating after it, it is clearly impossible to overtake it. No material object can exceed the speed of light, however much energy it has available. Mechanically, a body appears to grow progressively more ponderous as it is boosted faster. For example, subatomic particles which, in the reference frame of the laboratory, seem to be travelling at 99.9 per cent of the speed of light, are measured to be 22 times heavier than when at rest. As they are pushed towards the light barrier, their masses rise without limit.

This ability to convert energy into excess mass can be reversed – mass can be converted to energy. This happens in nuclear reactions, especially in the Sun, where four million tonnes of mass disappear each second to supply sunlight. The interconversion is expressed by Einstein's famous relation

$$E = mc^2.$$

The factor $c^2$ indicates that a little mass is worth a lot of energy. Just one kilogram of mass would power the average household for millions of years.

The fact that lengths and time intervals are dependent on the state of motion of the observer might appear to introduce a lot of complications into physics, but this is not so. Consider two events happening at two different places and two separate moments. The spatial distance $\Delta r$ between these events will depend on the observer, and so will the time interval $\Delta t$. However, it turns out that the combination

$$\Delta r^2 - c^2 \Delta t^2$$

is invariant, i.e., independent of the observer's motion. This is reminiscent of ordinary three-dimensional geometry, where the projection of a rod in three perpendicular directions, $\Delta x$, $\Delta y$ and $\Delta z$, is observer-dependent, varying according to the orientation of the rod relative to the observer, but the 'true' length

$$\Delta r = (\Delta x^2 + \Delta y^2 + \Delta z^2)^{\frac{1}{2}}$$

is constant, the same for all observers. Now we see that when motion is taken into account, we must also allow for length contraction effects from Einstein's theory of relativity, and the true invariant is not $\Delta r$ but

$$(\Delta r^2 - c^2 \Delta t^2)^{\frac{1}{2}} = (\Delta x^2 + \Delta y^2 + \Delta z^2 - c^2 \Delta t^2)^{\frac{1}{2}}$$

Thus, instead of combining the three spatial intervals in the usual Pythagorean way, we must now combine *four* intervals, $\Delta x$, $\Delta y$, $\Delta z$ and $c\Delta t$, although the latter is a bit peculiar because it comes into the invariant interval with a negative sign.

This type of space and time geometry replaces the old idea of three-dimensional space with a new four-dimensional *spacetime*, involving $x$, $y$, $z$ and $t$. Although separately space and time are relative, depending on the motion of observers, spacetime is absolute and independent of motion.

Fig. 1.15 depicts a region of spacetime. One dimension of space is suppressed for ease of drawing: time runs vertically, and each horizontal slice of the diagram represents space at one instant. The wavy line is the path in spacetime of a particle. This line is a history of the particle as it moves about, and is known as the particle's *world line*.

It is usual to use a scale of distance so that the speed of light is unity. This entails measuring time in, say, seconds and spatial distances in light seconds. The world line of a light pulse is then at 45° to the vertical in the spacetime diagram. If a flash of light is

Fig. 1.15. Spacetime diagram, showing a particle world line and forward light cone at the point $P$.

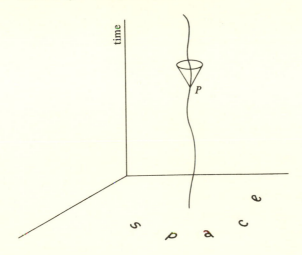

emitted in all directions by the particle at some instant (event $P$ in the figure) then all the world lines of light lie along a cone directed towards the future. This is called the *light cone*. Along this cone, velocity $= \Delta r / \Delta t = $ constant $= c$, so

$$(\Delta r^2 - c^2 \, \Delta t^2)^{\frac{1}{2}} = 0,$$

and we see that the four-dimensional distance along the light cone is zero. This value is invariant, so the light cone is agreed by all observers, although the world line of the material particle will change slope depending on which observer's reference frame is used. Physically this corresponds to the speed of light being the same in all reference frames, although the speeds of material objects are not.

The use of relativity and the four-dimensional spacetime language enables an elegant synthesis of electric and magnetic fields to be undertaken. The electromagnetic field has so far been described as a conjunction of electric field vector **E** and magnetic field vector **B** at each point of space. This looks like a sort of di-vector field – a rather complicated entity. However, it must be remembered that **E** and **B** are not completely independent of each other, but are coupled together via Maxwell's equations. To see what this means physically consider a familiar example of the relativity of motion. A flowing river constitutes a current to an observer standing on the bank side, but to an observer on an unpowered boat cut adrift on the stream, the water does not flow past the hull at all and there is no current.

Similarly, one observer may see a collection of charges sweep past and call it an electric current, while another, at rest relative to the charges, sees no current. The former will detect a magnetic field, but the latter will measure only a pure electrostatic field. One cannot say that there 'really' is a magnetic field or 'really' only an electrostatic field. The electric or magnetic nature of the system is relative to the state of motion of the observer, and only the synthesized electromagnetic field, considered as a whole, is independently 'real'.

Although a static electric field is a vector field (because it describes a force at each point in space, and force is a vector), nevertheless we do not need to specify **E** at each place to describe the field completely. This is because the value of **E** at one place is related to its value in neighbouring parts by Coulomb's equation. In fact, it suffices to give only a single number (i.e., a scalar, not a vector) at each point of space. This scalar is called the *electric potential*, and can be thought of as a measure of the electrostatic energy acquired by a unit test charge brought to that place. The static electric field is therefore really only a scalar field. To recover the direction of the electric vector from the corresponding scalar field we take the potential gradient. That is, at each point in space, the electric force points in the direction of greatest change (decrease) in the potential. The strength of the field is proportional to the gradient.

The fact that the electrostatic field is really only a scalar field is connected with the fact that it is generated by a scalar source – the electric charge density. In contrast, a magnetic field is generated by a *vector* source – the electric current. We must specify not only the strength of the current in a wire, but also the orientation of the wire before we can determine the magnetic field around it: current has both magnitude and direction. In the absence of free magnetic charge, the only time a magnetic field can be reduced to a scalar potential field is in a region of space away from current sources. If we wish to relate the field to the presence of a current, we must use a vector potential. The magnetic field is, therefore, in general a vector field.

In relativity, the status of vectors changes. Because of length contraction, an ordinary vector can appear to alter its magnitude from one reference frame to another (the 'arrow shrinks'). However, one can build four-component spacetime vectors in place of three-component space vectors, and these behave in an orderly way when the reference frame changes. That the electromagnetic field can be recast in this four-component language is no surprise, because electromagnetism was the midwife of relativity. What happens is that the electric potential (one-component, scalar) combines with the three components of the magnetic potential to form a four-component vector field with proper relativistic transformation properties. Therefore, we speak of the electromagnetic field as a four-vector field, or vector field for short, rather than a (three-dimensional) di-vector field.

# 2 What are gravity waves?

Electromagnetic waves are easy to understand, partly because they are familiar, partly because electric and magnetic forces have a simple, readily visualizable interpretation in terms of fields. Gravity, on the other hand, requires not only the theory of relativity, but an extension of that theory involving exotic notions like curved spacetime. This so-called general theory of relativity has a reputation for difficulty. Fortunately, an understanding of gravity waves can be achieved without the full extent of this theory, and everything the reader will need to know is explained in this chapter. In many ways, gravity waves behave similarly to electromagnetic waves, but there are differences too, and caution should be exercised in stretching the analogy too far.

## 2.1    Gravity as a force

Gravity is the most familiar force because it keeps our feet on the ground. It is also the most universal force, operating between all material bodies. We know that distant galaxies gravitate because their stars are bound to them by an invisible force. The stars themselves are held together against their enormous heat pressure by gravity. The gravity of the Moon (and Sun) is apparent on Earth in the daily tides.

The first systematic study of gravity was undertaken in the Middle Ages by Galileo Galilei and Isaac Newton. Galileo demonstrated that, ignoring complications such as air resistance, all bodies accelerate equally rapidly when they are dropped, irrespective of their mass or constitution.

Newton treated gravity as a force that acts at a distance across the empty space between material bodies. His celebrated laws of motion dictate that a body, if left to itself, will not alter its state of motion. Only if a force acts upon it will it change its speed or direction of motion. In free space, well away from other matter, a body will remain moving in a uniform way. On Earth, however, friction and air resistance soon sap the energy from moving bodies and bring them to a stop. Without such forces they would continue to move for ever.

Newton showed that the rate at which a body accelerates under the action of a force is inversely proportional to the quantity of matter it contains, i.e., its mass. The same force will accelerate a two-tonne truck half as rapidly as a one-tonne truck. The sluggishness of massive bodies resisting changes in motion is called inertia. If all bodies fall equally fast, therefore, the force of gravity must be stronger on the more massive ones in order to shift them equally effectively against their greater inertia. We know from experience that this is the case: massive bodies with great inertia are also heavy.

Physicists believe that Galileo's observation about falling bodies is *exact*. The force of gravity is *exactly* proportional to the mass of the body, i.e., the weight is exactly proportional to the inertia. It is a relation of profound significance, as we shall see.

Near the Earth's surface, gravity does not vary much, but an elementary observation shows that gravity must diminish with distance. The planets of the Solar System orbit the Sun in nearly circular paths at various distances, but the inner planets such as Mercury and Venus have much shorter orbital periods than the outer planets. Compare Mercury's 88 days with Jupiter's 12 years. Johannes Kepler discovered the following law for the planetary periods:

$$\text{orbital period} \propto (\text{orbital radius})^{\frac{3}{2}}.$$

Each planet avoids falling into the Sun by balancing the Sun's gravitational force on it with centrifugal force due to the planet's orbital motion. The latter is proportional to (orbital radius)/(period)$^2$, so using Kepler's law we immediately deduce that the centrifugal force on a planet is proportional to $1/r^2$, where $r$ is the orbital radius. To balance this, gravity must be an inverse square law of force – identical to that around an electric charge. The role of 'charge' is here played by the mass of the body, so Newton's law of gravity may be written

$$\text{Force} = \frac{GMm}{r^2}, \qquad (2.1)$$

where $M$ and $m$ are the two gravitating masses (idealized as points) and $r$ the distance between them. $G$ has the value $6.67 \times 10^{-11}$ N m$^2$ kg$^{-2}$. It determines the strength of gravity between two standard masses with standard separation. So far as we can tell, $G$ is a universal constant, which means that the strength of gravity is the same throughout the Universe and at all times.

If $M$ is the mass of the Sun and $m$ the much smaller planetary mass, $r$ is then the radius of the planet's orbit and we may equate (2.1) to the centrifugal force on the latter:

$$\frac{GMm}{r^2} = \frac{mv^2}{r} \qquad (2.2)$$

where $v$ is the speed of the planet. Solving (2.2),

$$v = \left(\frac{GM}{r}\right)^{\frac{1}{2}}. \qquad (2.3)$$

The speed of the outer planets (large $r$) is clearly less than that of the inner planets. The period is simply

$$\frac{\text{orbital distance}}{\text{speed}} = \frac{2\pi r}{v} = 2\pi \left(\frac{r^3}{GM}\right)^{\frac{1}{2}}, \qquad (2.4)$$

i.e., the square of the period is proportional to the cube of the orbital radius, which is Kepler's law.

The energy of a planet is made up of two components: its kinetic energy, $\frac{1}{2}mv^2$, and the gravitational binding energy due to its attraction to the Sun, $-GMm/r$. The latter is negative because one would have to do work to pull a planet away from its gravitational bond. The net energy is

$$\frac{1}{2}mv^2 - \frac{GMm}{r} = -\frac{1}{2}\frac{GMm}{r} \qquad (2.5)$$

using (2.3) to eliminate $v$. This result shows that the total energy is negative (bound orbit) and becomes more negative for planets near the Sun where the gravitational binding is stronger.

Although (2.5) has been derived for this idealized model with circular orbits, and the reaction of the gravitational force back on the Sun has been neglected (it is very small), nevertheless the structure of this result remains correct in general, even for complicated systems perhaps involving many masses. For a gravitationally bound system with total mass $M$, the total energy has the form

$$-\frac{1}{2}\frac{GM^2}{R} \tag{2.6}$$

where $R$ is a characteristic radius of the system (such as the mass-weighted average distance of each component from the centre of mass).

The fact that the centrifugal force of rotation can be used to cancel the Sun's pull of gravity is highly significant. Centrifugal force is sometimes called 'artificial gravity' because it is indistinguishable from the 'real thing'. Not only centrifugal effects, but any acceleration will do to simulate gravity (see Fig. 2.1). Conversely, if a body is allowed to accelerate freely under the action of gravity (think of a falling elevator) then the downward acceleration just cancels the pull of gravity and the system becomes weightless. This weightlessness is a familiar feature of spaceflight and occurs, *not* because an orbiting spacecraft is too far away to feel the Earth's gravity (or it could not be held in orbit), but because it is freely falling.

It sometimes causes confusion that an orbiting spacecraft is freely falling, for it does not drop to the ground. The reason for this is that, although the Earth pulls the spacecraft groundwards, there is also the orbital motion, which swings the craft to one side causing it to fall *round* the Earth rather than towards it (see Fig. 2.2). The

Fig. 2.1. Direct gravity is (locally) experienced in a way that is only relative to one's state of motion. Both observers feel the same $g$ forces – the one on Earth, the other on the accelerated platform in deep space.

Earth

acceleration is always towards the ground, but the motion is roughly perpendicular to this. Likewise, the Earth and Moon fall around each other's common centre of gravity, as do the Sun and planets.

Of course, these observations about weightlessness in free fall are really the same as those of Newton and Galileo that all bodies fall equally fast. When, for example, an elevator is cut loose, the framework and all its contents drop with the same motion, so an object released at the centre of the elevator will remain there – apparently weightless and suspended in 'mid-air' – as the whole assembly falls together (see Fig. 2.3).

The equivalence of gravity and acceleration was elevated to a

Fig. 2.2. Orbiting means falling. (*a*) Body *A* is dropped vertically from a point *P* in space and falls directly groundward. *B* is projected to the right at *P*, so falls to ground on a curved path. *C* is projected so fast to the right that it falls right past the Earth, overshoots, and falls back, misses again, and goes on falling round and round. (*b*) If the Earth's gravity were switched off, the projected body *C* would travel in a straight line to *C*$_1$. Instead the Earth accelerates it groundwards – it *falls* to position *C*$_2$ – so the orbit curves. Notice that *C accelerates* towards the Earth, but moves roughly perpendicular to this direction.

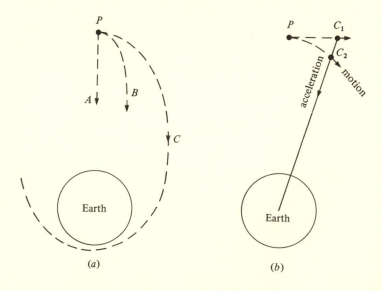

(*a*)                    (*b*)

'principle of equivalence' by Einstein, who was interested in how a change of reference frame (e.g., unaccelerated to accelerated frame) alters our experience of gravity. In section 2.3 we shall see how this equivalence suggests that gravity is not really a force at all, but a property of the space and time through which bodies move.

The fact that all objects fall equally rapidly under gravity means that an external gravitational force will not be noticed from any *local* observation, for everything will fall in the same way and no change will occur in the relation between nearby things. Thus the Sun's gravity, at the Earth's surface, which is actually as much as 0.06 per cent of the Earth's gravity, has no local effect at all: this amount of change would be easily noticed on any reasonably accurate spring weighing machine. We do not feel any lighter when the Sun is overhead. Nor should we notice it standing on the planet Mercury, at the surface of which the Sun's gravity is about 1 per

Fig. 2.3. Freely-falling frame of reference. Local gravitational effects vanish (are 'transformed away') in free all because of the principle of equivalence. All contents of the box fall downwards equally rapidly, so appear weightless to the observer. The bullet passing through the box appears to travel along a straight trajectory.

cent of that due to the planet itself. In both cases the respective planet is freely falling in the Sun's gravity, cancelling it completely by the orbital acceleration.

Although there is no local effect from gravity when the observer falls freely, there can be non-local effects. Return to the example of the falling elevator. Suppose two small bodies are released in the

Fig. 2.4. (*a*) Two nearby freely-falling particles slowly approach as they drop towards the geometrical centre of the Earth on gradually converging trajectories. (*b*) In this exaggerated view, four particles arranged in a square fall freely. Differential gravitational forces gradually distort the square to a diamond shape. The bottom particle is nearer the Earth, so feels a slightly stronger gravity and falls more rapidly. The top particle lags behind all the others as it is furthest from the Earth. The outside pair fall slowly together as explained in (*a*). These differential forces are called 'tidal' because they are also responsible for the ocean tides. (*c*) Similarly a flexible ring gradually becomes flattened at the sides and stretched top and bottom.

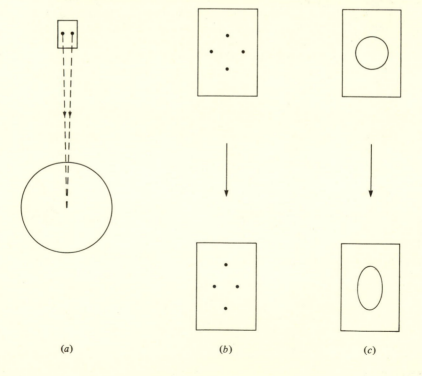

(*a*)          (*b*)          (*c*)

elevator some distance apart. Each falls directly towards the ground. However, the Earth is not flat, but approximately spherical, so the local horizontal beneath one particle is angled slightly relative to the other. Their vertical paths are therefore not exactly parallel, but slowly convergent, aiming towards the Earth's geometric centre (see Fig. 2.4). A falling observer would (in principle) notice a very slight motion of the two bodies towards one another as they plunged groundwards. In a real elevator, where the particles might only be separated by a metre, the convergence would only amount to a minute 0.016 mm for each vertical 100 m dropped. Nevertheless, it is a direct sign that an external gravitational force (the Earth's) is present.

Of course, in addition to these forces, there will also be minute gravitational forces acting between the falling particles. These internal forces will in general be negligible, and in our considerations here they will indeed be neglected. It is most important to understand that these relative movements between particles in free fall is *not* caused by any mutual gravity acting between them, but by their response to the *external* gravity.

The appearance of real effects even when freely falling under gravity is a non-local phenomenon because it requires observations of bodies spread over an extended region. The effects are produced if the external gravity is not uniform, i.e., if it varies in strength or direction from place to place. We are therefore dealing with a *differential* process – the direct gravity force itself is not observable (in free fall), but the usually much smaller secondary effect caused by its variation from place to place can be observable.

The larger the region of observation, the greater these differential effects can be. For example, the Moon's gravity at one location on the Earth does not exert a net force because the Earth and Moon fall freely in each other's gravity. However, the *variation* in the Moon's gravity across the Earth's surface is noticeable. Because the side of the Earth that is momentarily facing the Moon is more than 12 000 km closer to the Moon than the remote side, the Moon's gravity differs by over 6 per cent. This causes the oceans on different parts of the Earth's surface to fall at different rates towards the Moon, with the effect that tides are raised (see Fig. 2.5). The differential forces of gravity are therefore known as *tidal forces*. Note that although the Earth tries to fall towards the Moon it also has to contend with the Earth–Moon orbital motion, so the Moon and Earth end up falling *round* each other.

A striking illustration of the difference between direct gravity and tidal gravity is provided by the observation that, although the Sun's gravity at the Earth's surface is about 180 times stronger than that of the Moon, yet solar tides are smaller than lunar tides. This is because the *variation* in solar gravity across the Earth is only 0.017 per cent. Owing to the fact that the Sun is so much further away than the Moon, the additional Earth-diameter makes virtually no difference to the strength of the Sun's gravity.

Another illustration of the difference is provided by the fact that, although the Moon's gravity is weakest at point *B* in Fig. 2.5, nevertheless the ocean is raised there. This is because the Earth itself is almost rigid and so does not distort in shape. Thus, the

Fig. 2.5. Ocean tides. Just as the flexible ring in Fig. 2.4 (*c*) becomes flattened and stretched when it falls in the Earth's gravity, so the spherical shell of ocean around the Earth becomes flattened in a similar shape as it falls in the Moon's non-uniform gravity. The moon raises water at *A* and *B* and depresses it at *C* and *D*. The strength of the Moon's gravity at *A* is over 6 per cent greater than at the more distant point *B*, and it is this *variation*, not the direct gravity itself, that raises the tides.

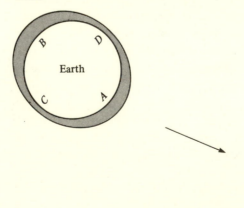

Earth's surface at $B$ falls at the same rate as the Earth's centre of gravity, and this is somewhat nearer the Moon than $B$. Hence the Earth at $B$ falls a little faster than the ocean at $B$, which is not rigidly attached to the Earth's centre of gravity. This means that the ocean is 'left behind', and a bulge occurs.

If the Earth itself were liquid, it would also deform in shape under the action of these tidal forces, and we should not notice the ocean tides locally. Because it is quasi-rigid, the tides induce *stresses* in the material. This tidal stress, attempting to strain the shape of a rigid body, will turn out to play an important role in the detection of gravity waves.

## 2.2    Gravity as a field

The force of gravity is similar in some respects to the force of electricity; for example, both obey an inverse square law. This makes it tempting to utilize the powerful concept of a *field* to describe gravity, by analogy with the electric field. However, what sort of field should we employ? What are its features?

There are important differences between electricity and gravity. Gravity is only known to attract bodies, whereas electricity can both attract and repel. This is because electric charge comes in two varieties: positive and negative. The role of 'gravitational charge' is played by mass, which is believed to be positive for all ordinary material bodies. Notice that with electricity, like charges repel, but with gravity, like masses attract.

Another difference between gravity and electricity is the extreme weakness of the former compared with the latter. The constituents of a hydrogen atom, for example, attract with electric forces about $10^{40}$ time more strongly than with gravity; if hydrogen were bound together by gravity, the cohesion would be so feeble that the smallest size of the atom (radius of the lowest quantum orbit) would be larger than the observable Universe! Nevertheless, the gravitational force between large metal spheres was measured in the laboratory by Henry Cavendish as long ago as 1798. Between two 1 kg spheres 1 m apart there is an attraction of only $6.7 \times 10^{-11}$ N. As gravity is so weak, only masses of astronomical size can exert appreciable forces.

A final difference is the fact that gravity can be *locally* simulated or abolished by acceleration, such as by free fall. This cannot be done with electric forces because not all bodies experience the same electric force. Indeed, some bodies (uncharged ones) do not respond to electric forces at all. If we release a collection of bodies with varying charges in an external electric field they will not behave in a 'weightless' fashion, all gliding along together, but will become rapidly dispersed as some are attracted more strongly than others, while some are even repelled owing to the sign of their charge.

There is a sense in which only tidal gravity can be considered as *real*. If direct gravity, i.e., local gravitational forces, can be simulated or abolished by acceleration, then the forces experienced are more a feature of the observer's state of motion than of gravity itself. We feel the Earth's gravity not just because we are situated in its gravitational field, but because the ground pressing on our feet restrains us from assuming a free-fall state of motion. An equivalent effect could be achieved in deep space, away from planetary gravity, by a $1g$ acceleration from a rocket motor (see Fig. 2.1). Only the absence of minute tidal forces in the spacecraft would betray the difference. All this suggests that any concept of a gravitational field should be based on the tidal effects, not the direct force. It also suggests that any waves in the gravitational field will be 'tidal waves'. The principle of equivalence of gravity and acceleration, with its implication of free-fall weightlessness, means that *objective* (as opposed to observer-dependent) gravity is a secondary, tidal effect.

If direct gravity is purely relative to one's state of non-uniform motion, it seems natural to extend the ideas of the relativity of uniform motion to the case of accelerated reference frames, in an attempt to incorporate gravity in the theory of relativity. Einstein undertook this task between 1905 and 1915, in which the earlier work, now called the 'special' theory, was expanded into the so-called general theory of relativity. Recall that special relativity is closely associated with the structure of space and time, or space-time, and that space and time individually can suffer distortions due to the uniform motion of observers. We shall see that Einstein's great discovery was that gravity can also be modelled by allowing spacetime to distort in a more elaborate way.

When a test charge is acted upon by an electric field, the resulting force can be represented by a single vector (see Fig. 2.6). In three dimensions this vector can be decomposed into three perpendicular components. However, the effect of tidal gravity is more complicated. Suppose we use as a test body a small cube of matter, and allow it to fall freely in a non-uniform gravitational field, such as the interior of our elevator plunging groundwards. In this simple example, the base of the block tries to fall faster than the top because it is nearer to the ground where the Earth's gravity is slightly stronger (recall Fig. 2.4). Tidal gravity therefore tries to stretch it vertically. In addition, the sides of the block try to fall on converging trajectories towards the centre of the Earth, so tidal gravity tries to squeeze it horizontally. In a more complicated gravitational field there might also be forces trying to twist the block, and to shear it away from rectangular form.

The combination of all these stresses acting on the block can be described by nine separate force components. Consider, for example, one face of the block. There will be forces parallel to the surface trying to shear and rotate it relative to the opposite face. In this two-dimensional surface, only two vector components are necessary to describe the net force. The forces acting on the opposing face will be equal and opposite, in order that no net force acts to move the centre of mass of the cube, for this would

Fig. 2.6. (a) An electric force can be represented by a single vector.
(b) Direct gravity is also a single vector, but tidal gravity acting on a freely falling block is more complicated. In each surface, forces try to rotate and deform the shape of the block. Other forces act perpendicular to the faces to try to stretch or shrink the block.

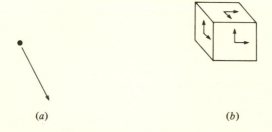

(a)                                              (b)

represent direct gravity, not just tidal forces. There are thus three independent pairs of vector components, one pair for each doublet of opposing faces. In addition, there will be equal and opposite forces acting perpendicular to each pair of faces, attempting to dilate or shrink the block. In all there are nine vector components of force (though in Einstein's theory only six are independent). This is in contrast to the three vector components of electric force.

It is clear that gravity is not simply a vector field, as with electromagnetism. The nine components necessary to specify the action of gravity on a freely-falling test body indicate a more complicated field structure. However, the nine components are not just random numbers. In fact, there is a sense in which the nine components can sometimes be treated as the conjunction of *two* three-components vectors (see box). This organization is expressed by calling this particular, rather special, nine-component mathematical object a *tensor* (strictly a tensor of the second rank). We therefore say that gravity is a tensor field whereas, as mentioned at the end of section 1.5, the electromagnetic field is a vector field. This distinction is most important.

In the theory of relativity, one deals not with three-dimensional space, but four-dimensional spacetime. Just as electromagnetism requires a four-vector field, so relativistic gravity will require a $(4 \times 4 =)$ 16-component tensor (though in Einstein's theory, only 10 are independent). The physical meaning of the additional components will be dealt with in due course.

If instead of a rigid body we drop a loose collection of particles in a cloud, then the aggregation will be unable to support tidal-induced stresses as they fall. The cloud will instead undergo a deformation of shape, just as the oceans change shape in the Moon's tidal gravity. The description of this deformation as shear, dilation and rotation will also require a multicomponent tensor to describe it. The changing shape can be regarded as small mutual accelerations of the individual particles relative to one another. The magnitude of these accelerations is proportional to the tidal force.

## 2.3    Curved spacetime

Although the distortion in shape of a falling collection of particles can be regarded as induced by tidal forces, it is possible to

# Gravity is a tensor field

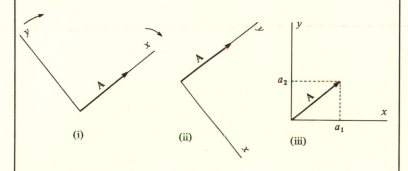

(i)   (ii)   (iii)

A is a vector of unit length. To describe it we could invent a coordinate frame $x$, $y$ and say (i) $\mathbf{A} = (1, 0)$. If this frame of reference is rotated, (ii) $\mathbf{A} = (0, 1)$. Intermediate orientations give intermediate number pairs, for example, (iii) ($a_1$, $a_2$). In all cases, $a_1^2 + a_2^2 = 1$. The two components $a_1$, $a_2$ must retain this relationship under changes of reference frame, to preserve the unit length of $\mathbf{A}$, so the components are not random numbers, but organized around this relationship. In three dimensions, three components ($a_1$, $a_2$, $a_3$) are needed: $a_1^2 + a_2^2 + a_3^2 = 1$.

Now consider two unit vectors $\mathbf{A}$ and $\mathbf{B}$, and multiply all their components pairwise: $a_1 b_1$, $a_1 b_2$, $a_3 b_2$, etc. Give the symbol $c_{11}$ to the pair $a_1 b_1$, $c_{32}$ to $a_3 b_2$, etc., and arrange all nine $c$s in a pattern as follows:

$$
\begin{array}{ccc}
c_{11} & c_{12} & c_{13} \\
c_{12} & c_{22} & c_{23} \\
c_{31} & c_{32} & c_{33}
\end{array}
$$

This nine-component object $\mathbf{C}$ is called a dyad, and is really a conjunction of the two vectors $\mathbf{A}$ and $\mathbf{B}$; one could write this as $\mathbf{C} = \mathbf{AB}$. As the reference frame is rotated, these components $c_{11}$, etc., change, but in a disciplined way that makes them more than just a matrix of numbers, for they must comply with the fact that *both* the lengths of $\mathbf{A}$ and $\mathbf{B}$, as well as their relative orientation, remain fixed. Any matrix $\mathbf{C}$ (whether constructed from two vectors or not) that enjoys these special transformation properties is called a *tensor*.

view the phenomenon in an entirely different and much more natural way. The notion of force is useful if we are dealing with a fixed force acting upon a collection of differing masses which undergo a variety of accelerations. With tidal gravity we have just the opposite. It is the accelerations that are fixed, while the tidal forces vary. This is because of the principle of equivalence, which tells us that all test bodies, whatever their mass or constitution, will accelerate equally in the tides. The tidal forces, of course, vary from particle to particle depending on the mass, but the acceleration at any given place is the same if we replace one test mass by a different one.

To emphasize this point, suppose we release in our falling elevator two rings of particles, one consisting of puff-balls, the other lead bearings (see Fig. 2.4). Then the effect of tidal gravity as they fall is to deform the ring shapes in identical fashion.

If gravity deforms all freely-falling shapes in identical ways, it suggests that a description of the deformation in terms of forces is unnecessarily complicated. The deformation is really only a *geometrical* effect, not a mechanical one. As it is the same for every type of test particle, we provide a more economical description of tidal gravity by casting the phenomenon in geometrical language instead of using the ideas of mechanics, and abandoning the use of forces altogether.

This was the approach adopted by Einstein. He argued that gravity was better understood not as a force at all, but as a manifestation of spacetime geometry. The deformation of the ring, for example, can be regarded not as a distortion of the ring itself due to forces, but a force-free tumble through a distorted geometry.

This type of situation is familiar in daily life. Suppose two golf balls are putted on parallel paths (see Fig. 2.7). So long as the green remains flat they will not alter position relative to each other. This is analogous to two nearby particles in a falling elevator where tidal forces are absent (uniform gravity). If the green is not flat, however, the balls may be deflected apart, or together, analogous to tidal forces altering their separation. Thus, a distortion from flat geometry can cause neighbouring particles to drift relative to each other, just as though tidal forces were acting.

Einstein proposed that spacetime is bent or distorted, and that as the particles fall, they travel along the straightest route, called a geodesic, through the curved background spacetime. Neighbouring geodesics can converge or diverge, reproducing the effect of tidal forces.

There is a good two-dimensional analogy to this idea. The surface of the Earth is curved, and this causes some oddities in air travel. To reach Los Angeles from London, one first flies northwest then southwest to keep to the geodesic ('straightest' or shortest route). The equivalent of a straight line on the Earth is a so-called great circle, like the lines of longitude. These particular geodesic paths all intersect at the poles. If two aircraft several kilometres apart fly north from the equator along their respective lines of longitude (geodesics) each pilot will regard his aircraft as travelling an exactly straight route. Although they start out flying parallel, they will slowly converge and eventually collide at the north pole (see Fig. 2.8). If the pilots believed in a flat Earth, they would have to explain this convergence as due to a mysterious external tidal force pulling them together. More sensibly, we now attribute the

Fig. 2.7. Balls moving on an uneven surface can experience forces that act like tidal gravity to cause them to drift together or apart. These forces can better be described as due to the distorted underlying geometry of the surfaces. (a) A hump in the ground causes the balls to drift apart. (b) A pit causes their paths to converge.

(a)

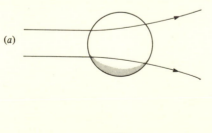

(b)

effect as due to the curvature of the Earth's surface. Similarly in the falling elevator, the superior description is that each particle is 'flying' the straight (geodesic) route, untroubled by any forces whatever, through a bent spacetime.

The two explanations of gravity as given by Newton and Einstein provide an interesting contrast. In Newton's theory, gravity is a force. The Earth, for example, moves in a curved orbit around the Sun because the Sun's gravity *forces* it away from its natural straight path. Einstein's theory describes the same phenomenon quite differently. The Sun's mass distorts the geometry of spacetime in its vicinity, and the Earth, gliding freely and without experiencing any forces, wanders along the straightest possible path in this bent background. The straightest (geodesic) path is roughly an ellipse, though with some slight deviations not predicted by Newton's theory, and which are actually observed. These give rise to the so-called perihelion precession of the planet Mercury.

It is important to realize that it is *spacetime* that is curved, not just space. There is frequently a misconception that the curvature of the Earth's orbit in space reflects the underlying curvature of space. This is not correct as a simple observation reveals. The Earth moves in a nearly circular orbit with a nearly constant speed. The curvature of its orbit in space is measured by $1/r^2$, where $r$ is the orbital radius; note that large $r$ corresponds to small curvature. For the Earth this orbital curvature is about $4 \times 10^{-23}$ m$^{-2}$. If another body, such as an asteroid, passes close to the Earth at

Fig. 2.8. Curved geodesics. Because the geometry of the Earth is not flat, two 'straightest' paths (geodesics) that are parallel at the equator, converge and intersect at the north pole. This is analogous to the tidal forces that cause two particles to drift together in a falling elevator.

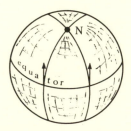

great speed, its orbit will be far less curved, even though it experiences the same gravity (see Fig. 2.9). In spite of this, when we go to *spacetime*, rather than just space, we see that these two orbits in fact have practically equal curvature.

To understand the meaning of curved spacetime one can study spacetime diagrams, introduced in section 1.5. Imagine that we could switch off the Sun's gravity; the Earth would then move through space with constant speed in a straight line. In Fig. 2.10 (*a*) this situation is depicted, with the world line of the Sun drawn broken to depict its gravitational impotence. The Earth's world line is therefore *straight* because the Earth is unaccelerated. Switch on the Sun's gravity (Fig. 2.10 (*b*)). The Earth now moves in *space* along a curved orbit that is reflected by the curvature in *spacetime* of its world line, which bends around the Sun's world line in a helix. If we project the helix down onto the space surface (i.e., view it from above) we see a nearly circular ellipse.

At this point we must remember that, when drawing spacetime diagrams, the scale along the space and time axes has to agree. This means multiplying time by *c*. Clearly, the diagram as drawn is very compressed vertically because the orbital period is one year (pitch of helix turns = one light year) whereas the orbital radius is only one astronomical unit ($1.5 \times 10^8$ km, or about $8\frac{1}{3}$ light minutes). To

Fig. 2.9. Curved paths do not mean curved space. At *P* the fast-moving asteroid intersects the Earth's orbit. Although both bodies are then equidistant from the Sun, the asteroid orbit is much less curved in space.

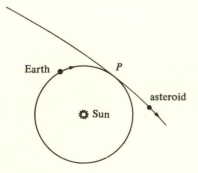

correct this compression, the drawing should be stretched out vertically by about ten thousand times. When this is done, the curvature of the Earth's world line in *spacetime* is very slight, though of course it is still $4 \times 10^{-23}$ m$^{-2}$ in *space*. Einstein proposed that this slight curving of the world line in spacetime is really due to the curvature of the spacetime geometry itself. This is rather as though we make a glass block model of spacetime containing straight world lines threading it, and then heat it and twist it about, so the lines are obliged to conform to the curvature of the glass.

Now it is easy to see that the high-speed asteroid on the shallow-curved orbit in space (see Fig. 2.9) has a world line in *spacetime* that curves about the same as that of the Earth. Although the

Fig. 2.10. Curvature in spacetime. (*a*) In the absence of gravity, the Earth flies off in a straight line. The path is straight both in space and spacetime. (*b*) With gravity the Earth is trapped in orbit around the Sun. In *spacetime* this orbit is a gently curving helix. To see its shape in space, view the helix from above (i.e., project it onto the horizontal plane) to obtain a much more strongly curved ellipse (dotted line). Remember that the vertical scale is very compressed, so the curvature of the helix is even less than depicted.

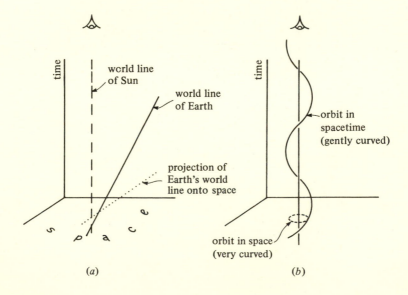

(*a*)                                          (*b*)

asteroid's trajectory in space does not bend so much, it travels one astronomical unit considerably more rapidly than does the Earth. Because the times involved, when multiplied by $c$, are so much greater than the distances, we may neglect the latter in calculating the curvature. Thus, if the Earth's period is $t$, in spacetime the orbital distance is about $ct$ (in this case one light year), so the curvature is about $16/c^2t^2$, or $16 \times 10^{-32}$ m$^{-2}$ for the Earth (see Fig. 2.11).

Compare this tiny spacetime curvature, equivalent to the slight bending of a circle one *light year* in radius, with the much greater space curvature for the Earth's orbit, which is a circle only one *astronomical unit* in radius. The spacetime curvature is a factor $10^9$ less.

Simple Newtonian mechanics (see Eq. (2.4)) gives the orbital period $t$ as $2\pi r^{\frac{3}{2}}/(GM)^{\frac{1}{2}}$ where $M$ is the mass of the Sun. Hence the

Fig. 2.11. To calculate the curvature of the helix in spacetime, we neglect the slight horizontal distance (radius of helix), which is very much less than the vertical distance (pitch of helix). The curvature $(1/r^2)$ is then approximately the same as that of a ci-circle, radius $r$, touching the helix, where $r \approx \frac{1}{4} ct$.

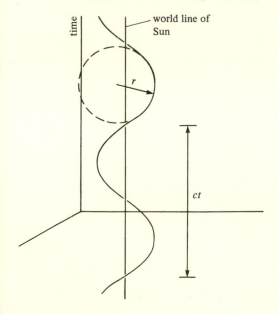

curvature of the Earth's world line due to the Sun is $16/c^2t^2 \approx GM/r^3c^2$. But $GM/r^3c^2$ is equal to

$$\frac{(GMm/r^2)}{mc^2} \times \frac{1}{r} = \left(\frac{\text{force of gravity}}{\text{mass}}\right) \times \left(\frac{1}{\text{orbital radius}}\right) \times c^{-2}$$

,

so

$$\left(\begin{array}{c}\text{curvature of} \\ \text{orbit in spacetime}\end{array}\right) = \left(\frac{\text{acceleration of gravity}}{\text{distance from Sun}}\right) \times c^{-2}. \quad (2.7)$$

However, the acceleration of gravity is (because of the equivalence principle) the *same* for the asteroid as the Earth, at the time that the asteroid passes the Earth's orbit. Moreover, its distance from the Sun is also the same as that of the Earth at that moment. Hence the spacetime curvature of the asteroid's orbit is, according to the above formula, the *same* as that of the Earth's. Both bodies, of course, are just reflecting the curvature of the underlying spacetime, which is about $GM/r^3c^2$. A proper treatment confirms this.

If two bodies move on slightly different orbits, they will accelerate relative to each other because of tidal effects (Sun's gravity varies from orbit to orbit). Rewriting (2.7) as

acceleration of gravity = (spacetime curvature) × (distance) × $c^2$

we note that, for two nearby particles,

$$\left(\begin{array}{c}\text{difference in} \\ \text{acceleration} \\ \text{due to gravity}\end{array}\right) \equiv \left(\begin{array}{c}\text{tidal gravity} \\ \text{acceleration}\end{array}\right) = \left(\begin{array}{c}\text{spacetime} \\ \text{curvature}\end{array}\right) \times \left(\begin{array}{c}\text{difference} \\ \text{in} \\ \text{distance}\end{array}\right) \times c^2$$

or

$$\left(\begin{array}{c}\text{tidal gravity} \\ \text{acceleration}\end{array}\right) = \left(\begin{array}{c}\text{spacetime} \\ \text{curvature}\end{array}\right) \times (\text{separation}) \times c^2. \quad (2.8)$$

## 2.4 The general theory of relativity

As remarked, the distortion of a falling flexible shape requires a multicomponent tensor to describe it. If we wish to translate this phenomenon into spacetime geometry, it is necessary to use a tensor from curved surface geometry, suitably generalized to four-dimensional spacetime. Fortunately, the mathematician Bernhard Riemann had already done most of the work, and, in his general theory of relativity, Einstein employed the so-called Riemann tensor as a description of spacetime curvature. The tidal effects can then be expressed schematically as

tidal acceleration vector components =
(Riemann tensor) × (separation).                                    (2.9)

The exact definition of the Riemann tensor need not concern us. It contains twenty independent components that encode all the geometrical information about how the spacetime curves in different directions. This result should be compared with (2.8). In the simple example considered there, only one component of the Riemann tensor was employed to estimate curvature. This was legitimate as we assumed the bodies were moving slowly, compared with light (non-relativistic approximation), and in circular orbits. Given the Riemann tensor, the motions of falling test particles can be computed in a routine way. The answers frequently agree almost exactly with Newton's old theory of gravity, but there are small corrections such as the one which explains the perihelion precession of the planet Mercury.

Einstein not only showed that gravity is a tensor field, he also discovered the *field equations* which relate the degree and nature of the spacetime distortion to the qualities of the gravitating material. Nothing has yet been said about the system that produces the gravity, and a full theory must take this into account also.

If gravity is a tensor field, it must have a tensor source. Newton's theory of gravity was based on the idea that the *mass* of a body is the source of gravity. This quantity is just a number (a scalar), not a tensor. With the discovery that energy is equivalent to mass ($E = mc^2$), it was clear that the source of gravity is something more complicated than mass. To see the further complexities, imagine a particle at rest. It has mass, but no additional energy. However, another observer moving relative to the particle will ascribe to it both rest mass and kinetic energy. Hence its gravitational effect is altered. The transformation of the gravitational field tensor under this change of reference frame must reflect these considerations.

Energy is a number and, as apparent from the above, it is one which depends on the observer's reference frame. This is analogous to electric potential (recall the discussion on page 25). In the latter case the theory was improved by building the non-relativistic scalar field into a properly-transforming four-vector, i.e., the electromagnetic potential. A similar manoeuvre works for energy by combining it with the three-vector *momentum*. The energy–momentum (or mass–energy–momentum) four-vector, is a physical quantity with correct transformation properties under a change of reference frame.

However, we still do not have a tensor: the mass–energy–momentum four-*vector* will not yet do as a source of gravity. A simple tensor could be made out of a conjunction of two such four-vectors (see box on page 39). That turns out to be close, but there are further requirements: the mathematical properties of the source tensor must correspond with the geometrical properties of the Riemann tensor and its related tensors. This makes it necessary to incorporate the effects of *stress*, as well as mass, energy and momentum. One ends up with the three-dimensional stress tensor augmented by the mass–energy–momentum four-vector. The latter makes up the additional components required for turning a three-tensor into a four-tensor, which can be represented by a $4 \times 4$ matrix (see page 38). In Einstein's theory only ten of these sixteen components are independent.

In the final field equation Einstein uses a tensor based on a contracted form of Riemann's tensor, having only ten independent components, as the geometrical description of the spacetime distortion, and equates it to the ten independent components of the stress–energy–momentum tensor: schematically,

variations in geometry of spacetime = stress, mass–energy and momentum of source.

This is the famous Einstein gravitational field equation. It plays a role analogous to Maxwell's equations (see page 18), which relate the four independent vector electromagnetic field components to the charge density and current vector of the source (also four components).

## 2.5    Gravity waves

The question immediately arises as to whether gravity waves can occur as solutions of Einstein's equation in the same way that electromagnetic waves are solutions to Maxwell's equations. As early as 1918 Einstein himself addressed this question. He did indeed find that wavelike solutions exist, in which an undulation of spacetime propagates through empty space as an independent entity, at a velocity equal to that of light.

In the early days there was some confusion about the status of these solutions. Some people argued that the ripples of geometry were not physically real because, like a direct (local) gravitational force, they could be transformed away by an acceleration, meaning that the travelling distortions were only a mathematical artifice describing a change of reference frame, like going from Cartesian to curvilinear coordinates on a still flat sheet of paper (see Fig. 2.12).

That there is real physics in the waves is clear when they are written in terms of the Riemann curvature tensor, which satisfies a wave equation virtually identical to the electromagnetic field. The Riemann tensor represents a real distortion of geometry, not just a change of coordinates.

But what *are* gravity waves?

Let us return to Fig. 1.11 and the explanation of electromagnetic waves in terms of the sudden relocation of a source particle. An analogous argument can be developed for gravity. If a massive

Fig. 2.12. Real and apparent curvature. (*a*) The flat sheet has a rectangular grid. (*b*) The same flat geometry is mapped by a complicated, curved grid. (*c*) This surface is intrinsically curved and no rectangular grid can be constructed. Early workers worried that gravity wave undulations were purely fictitious – mathematical waves of the grid (like (*b*)) – but they are in fact real ripples of geometry (like (*c*)).

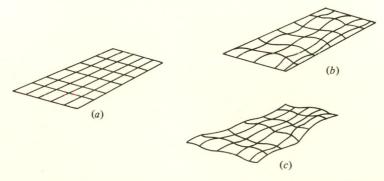

(*a*)

(*b*)

(*c*)

body is violently disturbed, the near field adjusts rapidly, but the far field must wait for the signal that the mass has moved to propagate out to it at the finite speed $c$, so there is a travelling kink which falls off in strength like $1/r$ rather than the $1/r^2$ Newtonian force.

There is one major difference, however, related to the fact that electricity can be both attractive and repulsive whereas gravity is purely attractive (electric charges are both positive and negative whereas mass is always positive). A simple electric dipole radiator consists of a positively charged particle attached to a spring anchored to a very heavy body (see Fig. 2.13). As the charge oscillates back and forth it constitutes a changing dipole, and electromagnetic waves are generated. The dipole moment is $ex$, where $x$ is the displacement and $e$ the charge. As the particle accelerates, so does the dipole moment. This arrangement can be replaced by *two* particles, one carrying positive charge, the other negative charge. When joined by a spring, the oscillations of this doublet constitute a dipole with similar characteristics to the single charged particle. The reason is that the displacement of a positive charge to the left is equivalent electrically to the displace-

Fig. 2.13. Electric dipoles. The oscillations of the single charge shown in (*a*) have the same electric dipole characteristics as the oscillating doublet shown in (*b*). In both cases when the spring contracts there is a net motion of positive charge leftwards, while an expansion shifts the positive charge rightwards. This oscillation of the 'centre of charge' generates electromagnetic waves.

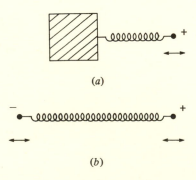

(*a*)

(*b*)

ment of a *negative* charge to the right. Consequently, when the spring contracts, the motion of the positive charge leftwards is reinforced by the negative charge moving rightwards; there is a net motion of (positive) charge to the left. When the spring expands there is a net shift of (positive) charge to the right. So the vibrating doublet is equivalent to (twice) a single oscillating charge.

Coming now to gravity waves, consider two equal masses joined by a spring (see Fig. 2.14). In this case, *both* 'gravitational charges' (i.e., masses) are positive, so the motion of one to the right is *not* equivalent to the motion of the other to the left. In fact, gravitationally, the two effects work to oppose rather than reinforce each other, so this doublet does *not* constitute a changing dipole at

Fig. 2.14. Gravitational dipoles cannot generate gravity waves. (*a*) Both masses carry positive gravitational charge, so as the spring contracts the leftward motion of one mass gravitationally opposes the rightward motion of the other. The centre of mass does not move, so the dipole does not oscillate. (*b*) This would constitute an oscillating dipole if the heavy mass were truly fixed. However, there is always a slight reaction motion and, because the heavy mass has a much greater gravitational action, it still cancels the gravitational dipole motion due to the light mass; the centre of mass still does not accelerate. We have to look to the quadrupole moments of these systems (which do oscillate) for the generation of gravity waves.

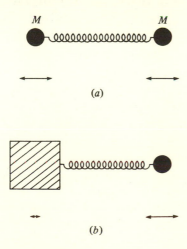

(*a*)

(*b*)

all. A mass dipole could consist of a *single* mass, after the fashion of Fig. 2.13 (*a*), but this time there is a problem. When considering the vibrations of the charged particle, the heavy body to which the spring was anchored could be ignored because it was uncharged. With gravity, however, there are no 'uncharged' bodies. All bodies have mass and therefore gravitational charge, so we must take the gravity of the anchor mass into account.

When the spring contracts it pulls the particle to the left, but the remote end also tries to pull the heavy body to the right (action and reaction are equal and opposite). Because the heavy body is more ponderous, its greater inertia prevents it from moving much. However, the gravity of this body is all the greater for its large mass. To reduce the motion of this anchor body, we can increase its mass, but there is a trade-off with increased gravitational action. Because of the equivalence principle the two effects exactly cancel. The effective gravitational disturbance is the same *whatever* the mass of the heavy body. So there is no way to make an oscillating mass dipole.

This result can be summarized by invoking the law of conservation of momentum. The centre of mass of the total system shown in Fig. 2.14 cannot accelerate, so the mass dipole moment cannot accelerate.

The double-mass systems shown in Fig. 2.14 in fact constitute a *quadrupole*, each individual mass acting as an oscillating dipole. To a first approximation the gravitational disturbance from one oscillating dipole cancels that of the other, but, because of the time required for the disturbance to propagate across the length of the spring, the two disturbances are slightly out of step and exact cancellation does not occur. The system radiates gravity waves, as a result of this changing quadrupole moment.

Just as the strength of the dipole moment governs the efficiency of generation of electromagnetic waves, so the strength of the quadrupole moment controls the output of gravity waves. In this simple case of two equal masses $M$ joined by a spring of length $x$ the quadrupole moment in the $x$-direction is

$$Q = Mx^2. \qquad (2.10)$$

There will be quadrupole moments in the perpendicular directions also. The general definition of $Q$ is

$$Q_{xx} = \int \rho \, (3x^2 - r^2) \, \mathrm{d}^3x \qquad (2.11)$$

for the $x$-direction, where $\rho$ is the mass density, $r^2 = x^2 + y^2 + z^2$ and $\mathrm{d}^3x$ means integration over the volume of the body. Similar expressions apply for $Q_{yy}$ and $Q_{zz}$. In general there may also be cross-components like $Q_{xy}$, defined by

$$Q_{xy} = 3 \int \rho \, xy \, \mathrm{d}^3x \qquad (2.12)$$

etc.

There is also a good geometrical reason why a quadrupole, rather than a dipole, is needed as the source of gravity waves. The electromagnetic field is a vector field, and so electromagnetic waves can be generated by vector sources, such as an electric current or the sideways motion of an electric charge. This means that a dipole source is sufficient (a dipole can be described by a vector). Gravity, on the other hand, is a tensor field, and the source must contain more components than a dipole (vector) to stimulate it. As remarked on page 38, a tensor may be regarded as the conjunction of two vectors, and so the source must be at least as complicated as two vectors. The simplest such arrangement is the quadrupole (see Fig. 1.13) consisting of two opposed vector dipoles. Thus, gravity waves will be generated if two nearby masses accelerate in opposite directions. The resulting field pattern will reflect this more complicated arrangement ($\sin^2\theta$ rather than $\sin \theta$ angular field dependence).

It can be shown that in the radiation zone (i.e., far from the source) the gravity wave also travels as a transverse vibration of the field. As with the electromagnetic field there are two independent polarization states of the wave (see below).

To understand the precise meaning to be attached to the gravity wave we may consider its effect on test particles placed in its path, just as the electromagnetic wave manifests itself by its effect on test charges. As the gravity wave is a tensor, we need a more complicated test system than just a single particle. A convenient configuration is to take a flexible ring placed perpendicular to the wave-propagation vector **k**. Figure 2.15 shows the physical effect induced by the passage of the wave. The ring distorts from circularity as the undulation passes. During the first half-cycle the distortion is in one direction, then in the second half-cycle it distorts in a perpendicular direction.

Geometrically, the reason for the distortion is that the gravitational wave represents a local change in the geometry as it passes, which stretches or shrinks all distances at right angles to each other. In more mechanical language, the wave represents the passage of a pulse of tidal gravity and the ring suffers the same sort of tidal rearrangement as the oceans of the Earth (compare also Fig. 2.4 (c)). The different parts of the ring fall different ways in the slightly different gravitational fields experienced at the various regions of its periphery.

Figure 2.15 (a) shows a gravity wave with one particular polarization. In the case of an electromagnetic wave, the other polarization state is rotated by 90° relative to the first, i.e., the electric fields of the two states are perpendicular (recall Fig. 1.8). Here a rotation through 90° of the configuration shown in Fig. 2.15 (a)

Fig. 2.15. Effect of gravity wave. The passage of the wave perpendicular to the plane of the flexible ring induces the distortions shown. Both polarization states are illustrated. The broken lines represent the distortions half a cycle later. The ring distortion therefore oscillates between perpendicular directions. Beneath each figure a single-cycle sequence of distortions is depicted.

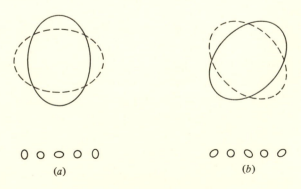

(a)                              (b)

reproduces the same pattern, but one half-cycle out of phase, so this is *not* an independent polarization state. Instead, the other state is rotated only 45°, as shown in Fig. 2.15 (*b*). Thus, a rotation of 90° takes one through twice as many gravity wave polarization states as electromagnetic ones. This is no surprise as the quadrupole source contains, roughly speaking, two vectors rather than one.

As in the electromagnetic case, any quadrupole wave configuration can be built out of a superposition of the two polarization states. One special case is obtained if they are set out of phase by one quarter-cycle. This is the case of circular polarization again (see page 12). The net effect on the flexible ring is to cause the bulges to rotate as the wave passes (Fig. 2.16). Note that after one complete wave cycle the pattern rotates through 180°, compared with 360° for the electromagnetic circularly polarized wave. It therefore takes *two* cycles for one complete rotation of the pattern, and this double-valuedness (see also above) derives from the tensor character of the wave.

It has an interesting consequence for the *spin* of the wave. If two geometrically identical bodies rotate with the same energy, but one has half the frequency of the other, then the former must carry twice the angular momentum (see Fig. 2.17). Similarly, if two waves have equal energy, but one rotates the test system at half the speed, it also carries twice the angular momentum. Thus gravity waves have a spin of twice that of electromagnetic waves. If we call the photon a spin 1 (i.e., $h/2\pi$) particle, then the quantum analogue for gravity – the 'graviton' (see section 5.4) – has spin 2 (i.e., $h/\pi$).

Fig. 2.16. Circular polarization. The distortion rotates rather than oscillates. Note that although the shape returns to the same configuration after only one cycle, it is 'upside down', and a complete revolution of the bulge requires *two* wave cycles. Contrast this 'spin 2' character with the 'spin 1' of electromagnetic waves (see Fig.1.9).

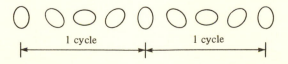

| 1 cycle | 1 cycle |

It is curious to note the similarity between the rotation of the bulges depicted in Fig. 2.16, with those in Fig. 2.5, i.e., the bulge of ocean tide which rolls around the Earth as the Moon revolves in its orbit. The periodic cycle of the Moon does indeed set up a gravity wave with period one month, but this is the *near field* ($1/r^2$ field, corresponding to the Coulomb force in electrodynamics) not the radiation, or far field ($1/r$ field). It is the latter gravitational *radiation* waves that are the subject of this book.

Before finishing this introduction to gravity waves, a serious theoretical complication must be mentioned. In Einstein's theory, gravity can be caused by stress, mass–energy and momentum. A gravitational wave carries energy and momentum, and will therefore itself act as a source of gravity. Because gravity represents energy, we may crudely say that gravity gravitates. The graviton is 'charged' with mass–energy. In contrast, the photon carries no electric charge, and is not a source of electromagnetic fields. This difference between electromagnetism and gravity is expressed by saying that the former is linear and the latter non-linear. A linear system generally has the property that, if two causes give rise to two effects, then both causes operating together produce the sum of the effects. For example, if one source produces a certain electromagnetic field and another produces a different field, then both

Fig. 2.17. The rings rotate with the same energy, but (*b*) spins twice as fast as (*a*). The angular momentum is defined as mass × angular velocity × radius² $\equiv M\omega r^2$. For rings of the same radius the ratio of angular momenta is $M_1\omega_1/M_2\omega_2$. But kinetic energy $=\frac{1}{2}M_1\omega_1{}^2r^2=\frac{1}{2}M_2\omega_2{}^2r^2$ so $M_1\omega_1/M_2\omega_2=\omega_2/\omega_1=2$. Thus (*a*) has twice the angular momentum of (*b*). (It also must have four times the mass.)

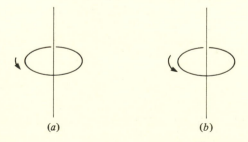

(*a*)　　　　　　　(*b*)

sources together produce both fields superimposed. Therefore, we may add together electromagnetic fields in a linear way – like ordinary numbers or vectors.

Gravity, being non-linear, is more complicated. If two masses individually produce associated fields, when they act together the field is more than the direct sum of the two. Account must be taken of such things as the gravity resulting from the gravitational interaction energy between the two sources, and the gravity of one field acting on the other.

Fortunately, if the fields concerned are weak, the non-linear effects are usually small and can be ignored. Frequently this is assumed in discussions of gravitational radiation. The wavelike disturbances that travel at the speed of light and behave as close analogues of electromagnetic waves are solutions of Einstein's equation which are being treated in this linear approximation.

# 3    Sources of gravity waves

Establishing the possibility of gravity waves is very different from proving that they are detectable or important. It took only a decade or so from when Maxwell predicted radio waves for them to be produced by Hertz in the laboratory. It is now over sixty years since Einstein's original paper on gravitational radiation. How close are we to generating these waves in the laboratory?

On page 35 it was mentioned that gravity is, in a sense, about $10^{40}$ times weaker than electromagnetism. This extreme feebleness is the major obstacle to the technological manipulation of gravity and almost certainly means that the study of gravitational radiation will have to rely on natural sources elsewhere in the universe. Unfortunately, the same feebleness that inhibits the generation of gravity waves also afflicts their detection, so that even a flux of gravitational radiation energy at the Earth's surface comparable to the heat and light of the Sun is unlikely to be detected by present terrestrial equipment. So weak is the interaction between gravity waves and matter that only one graviton in $10^{23}$ registers on a typical detector. This means that enormously powerful processes are necessary to render any practical interest to the subject of gravity waves. In this chapter some of these awesome processes will be described, and their gravitational luminosity assessed.

## 3.1    Simple generators

In principle the generation of gravity waves is extremely simple. All that is needed is a changing quadrupole of mass. For example, the device shown in Fig. 2.14(a) – two equal masses joined by a spring – would suffice.

Restricting the treatment to the linear approximation, the rate of energy emission from this system may be estimated by noting that the output will depend on the strength of gravity through Newton's constant $G$, on the speed with which the energy flows away from the system, which is the speed of light $c$, and of course on the quadrupole moment $Q$. One would also expect a dependence on the angular frequency of vibration $\omega$. These are the only physical parameters that characterize the system.

When set into vibration, the masses will oscillate about their equilibrium position. Their separation $x$ will therefore vibrate sinusoidally about the equilibrium length $L$:

$$x = L + x_0 \sin \omega t$$

where $x_0$ is the amplitude of the oscillations (assumed small). The relevant (i.e., oscillating) part of the quadrupole moment $Q = Mx^2$ is therefore $2MLx_0 \sin \omega t$.

Purely on dimensional grounds, the power rate averaged over one cycle must be given by the simple formula

$$\frac{\alpha M^2 L^2 x_0{}^2 \omega^6}{(c^5/G)}, \tag{3.1}$$

where $\alpha$ is a dimensionless constant. A full treatment using general relativity yields the value $\alpha = 1/15$. The universal constant $c^5/G$ is characteristic of gravitational radiation power output and will recur in all the power formulae in this chapter. It too has units of power, or energy flow per unit time

$$\frac{c^5}{G} = 3.6 \times 10^{52} \, \text{J s}^{-1}, \tag{3.2}$$

and represents a sort of power standard against which to gauge the efficiency of gravity wave production.

The remaining factor in (3.1), $M^2 L^2 x_0{}^2 \omega^6$, is variable, depending on the system concerned, and has units of (power)$^2$. The power output is therefore determined by the ratio of this quantity to the 'standard' power output (3.2). Notice that $c^5/G$ is a colossal energy rate – more than the heat and light output of the entire Universe! It corresponds to the conversion of 200 000 solar masses into energy every second. The presence of $c^5/G$ in the denominator of (3.1) indicates that unless the numerator involves energies of astronomical proportions, the value of (3.1) will be pitifully small. For two 1 kg masses, 1 m apart oscillating through 1 cm at 10 Hz (cycles per second), the power radiated is only $5 \times 10^{-47}$ J s$^{-1}$. It would need a thousand billion billion billion billion billion such devices to power a single domestic light bulb with gravity wave energy.

A more suggestive way of writing (3.1) is

$$\frac{\ddot{Q}_{xx}{}^2 + \ddot{Q}_{yy}{}^2 + \ddot{Q}_{zz}{}^2}{45(c^5/G)},\tag{3.3}$$

in terms of the third rate of change of the quadrupole moment $Q$ ($\dddot{Q} \equiv d^3Q/dt^3$). This formula turns out to be correct generally, not just for the simple system considered above. If there are cross-components in the quadrupole moment (e.g., $Q_{xy}$), their contributions are simply added to (3.3).

One efficient laboratory source of gravitational radiation would be a rotating bar. Seen from the plane of rotation, the projection of the bar expands and contracts periodically as it rotates between an end-on and a sideways-on configuration. This motion therefore amounts to an oscillating quadrupole (see Fig. 3.1). For a rod of

Fig. 3.1. Rotating bar. (*a*) This simple system is a source of gravity waves. (*b*) Seen from the plane of rotation, the bar appears alternately to expand and contract between these two shapes, thereby constituting an oscillating quadrupole moment.

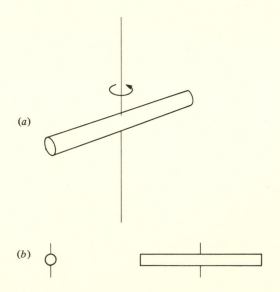

mass $M$ and length $L$ rotating $\omega$ times a second, the universal formula (3.3) yields

$$\frac{2}{45} \frac{M^2 L^4 \omega^6}{(c^5/G)}. \tag{3.4}$$

It is interesting to speculate how large this output could be made using present-day technology. The bar must not rotate too fast or it will snap. Using extremely tough material, such as steel, one could conceive of a huge lump weighing, say, 100 000 tonnes spinning at one revolution per second. This awesome device would generate about $10^{-24}$ $Js^{-1}$, still absurdly small compared with a domestic light bulb. Clearly, laboratory generated gravity waves are of little conceivable technological value.

The greatest output is obtained by large values of the mass and frequency. To see what is possible, it is instructive to rewrite (3.4) in terms of the rotation velocity $v$ of the periphery:

$$\frac{128}{45} \frac{(M/L)^2 v^6}{(c^5/G)}, \tag{3.5}$$

or, more suggestively,

$$\frac{32}{45} (r_g/L)^2 (v/c)^6 (c^5/G), \tag{3.6}$$

where

$$r_g = \frac{2GM}{c^2} \tag{3.7}$$

is a distance known as the *gravitational radius* of a mass $M$. If it were possible to obtain both the ratios $r_g/L$ and $v/c$ in the region of unity, the huge power output $c^5/G$ ($10^{52} Js^{-1}$) would be approached. Thus, very considerable gravity wave effects might occur in systems which are moving close to the speed of light ($v \approx c$) and are so compact that they approach their gravitational radii ($L \approx r_g$).

As a guide, note that the Earth rotates at $10^{-6}$ $c$ and has a radius some $10^9$ times its gravitational radius. However, as we shall see, there do exist astronomical objects that, at least temporarily, come close to satisfying the above requirements.

Another interesting way of writing (3.5) is

$$\frac{512\,\pi^2}{45}\left(\frac{G}{c^5}\right)\left(\frac{\frac{1}{2}Mv^2}{T}\right)^2,$$

where $T$ is the rotation period. This quantity is roughly

$$\frac{G}{c^5}\times\left(\frac{\text{kinetic energy}}{\text{time for energy to flow round system}}\right)^2,$$

or

$$\frac{(\text{power flow through system})^2}{(c^5/G)}$$

$$=\frac{(\text{system power})^2}{\text{'standard' power}}. \tag{3.8}$$

Although (3.8) was derived from the formula for the rotating bar, it is perfectly general, and also follows from (3.3) if we note that generally $Q\approx\text{mass}\times(\text{system size})^2$ (see definition of $Q$ on page 53. To illustrate the use of this formula, let us estimate the gravity wave power output from one of the more violent conceivable terrestrial events – the impact of a huge meteorite with the Earth. A rock $10^3\,$m in diameter striking solid material at a speed of $10^4\,\text{ms}^{-1}$ would set up a shock wave. At the moment of impact this disturbance represents a changing quadrupole. The time scale for wave propagation across $10^3$m is of order $0.1\,$s and the meteoric mass is about $5\times10^{12}\,$kg, so the effective $\mathrm{d}^3Q/\mathrm{d}t^3$ will be around $(5\times10^{12}\,\text{kg})\times(1000)^2/(0.1)^3$, giving a gravity wave power output for a fraction of a second of roughly

$$\frac{G}{c^5}\,(5\times10^{21})^2\approx10^{-9}\text{Js}^{-1},$$

which is still tiny by any standards, and far too small to be detected with present technology.

The only real hope for studying gravity waves is to look beyond the Earth to extraterrestrial processes of astronomical magnitude. However, our location in the Universe is of necessity well removed from the most violent and energetic systems (or life would not survive), so the best gravity wave generators are likely to be the most distant. There is thus a trade-off: the powerful sources are far away and only deposit a small fraction of their output on the Earth.

Fig. 3.2. Galaxy M81. This beautiful spiral structure is a typical galaxy, much like our Milky Way, containing about $10^{11}$ stars and measuring perhaps 100000 light years in diameter. Gravity waves are likely to come from black holes in the bright nucleus, or from stellar explosions in the spiral arms. Our Sun is situated in one of the spiral arms of the Milky Way, about a third of the way out from the centre. (*Lick Observatory, California*)

There is also a more serious snag. It is true that passing from laboratory, or even terrestrial, to astronomical masses boosts the power output by many orders of magnitude. However, inspection of the general formula (3.8) shows that it is the rate of energy *flow* that matters, not just the mass–energy itself, so the question has to be asked as to what forces are responsible for churning these astronomical masses about. The *rate* of mass rearrangement will depend on the nature and strength of these forces. When conspicuous astronomical objects are considered, gravity is usually the most dominant force. This means that the forces that cause violent events in massive astronomical systems are themselves gravitational, and it turns out, as we shall see, that gravitationally induced power flow has a severely depressing effect on gravity wave generation.

### 3.2 Efficiency of gravity wave generation

Even assembling enormous masses so that the gravitational charge can be boosted by many orders of magnitude does not guarantee the efficiency of gravitational wave production. The reason for this is that very massive objects (stars, planets, black holes, galaxies, etc.) are so ponderous that electromagnetic forces hardly affect them. Only gravity can move them around with any violence. However, a body accelerated by gravity itself is a very poor emitter of gravity waves.

To understand why this is so, recall the equivalence principle, which says that all matter and energy (including gravitational field energy) accelerates equally fast when it falls freely under gravity. This makes it very hard for a body to 'shake off' its own gravitational field as gravity waves, purely as a result of falling freely in some other system's gravity.

It is instructive to see this in detail. First let us see why the equivalent electromagnetic process is so efficient. Imagine a small electrically charged particle surrounded by an electric field which extends out to infinity, though the strength of the field diminishes with distance from the particle. If the particle is acclerated it will emit electromagnetic waves, as explained on page 14. Suppose the particle is accelerated by an external electric field, such as that due to another particle (this would be a collision between two charged particles – a common source of electromagnetic waves). The charged particle is deflected violently from its path by the external field, but its *own* field, which is uncharged and therefore does not feel a force – it is unaffected by the external field – tries to keep on going as before. Indeed, the distant regions of the field will not be aware of the sudden deviation for some time. Thus, the field shape is buckled and the distortion propagates away as electromagnetic radiation. In a sense, the edges of the field are stripped off when the source particle is jerked suddenly to one side. The effect is most pronounced very far from the particle, though of course the field there is correspondingly weaker.

Coming now to the gravitational case, let us first consider an *electric* particle accelerated by a gravitational field. Once again the charged particle is violently deflected in its motion as it falls in the external gravity, but this time, the electric field energy which surrounds it also feels the gravity, for energy is subject to gravity. Moreover, because of the equivalence principle, the field energy falls with an acceleration equal to the central particle, so at first sight it would appear that no electromagnetic radiation is produced by the falling charge – the particle plus electric field simply tumbles downwards, with the electric field shape undistorted.

However, it must be remembered that the equivalence principle is only true *locally* whereas the electric field around the charge clearly extends through space. Thus the periphery of the field will probe regions where the external gravity is different in strength from its value at the charged particle. In particular, if the external gravity is caused by a compact massive body, then the electric field far from the body will try to fall somewhat less slowly in the weaker gravity there, while the electric field close to the body will be jerked even harder downwards than the charged particle. So these tidal forces also try to pull the electric field from the particle, but considerably less vigorously than would an external *electric* field. The tidal violence distorts the electric field shape, and the disturbance propagates away as electromagnetic waves, i.e., some of the electric field 'keeps on going' as radiation while the charged particle falls away from within it. But the level of radiation is much lower than with electric acceleration because it is only a tidal effect.

These considerations can be extended, by direct analogy, to the case of gravitational waves. The falling body has mass and is surrounded by its own gravitational field which, being energetic, is also subject to the external field (remember, gravity gravitates), so tries to fall with the body. Once again tidal forces distort the field shape and strip off ripples of gravitational wave energy, but very inefficiently.

These heuristic descriptions of the (suppressed) production of electromagnetic and gravitational radiation by freely falling bodies receive also a very natural interpretation in terms of curved spacetime, for when the falling field encounters tidal gravity, it is really encountering curvature. The geometrical distortion then 'kinks' the force lines after the fashion of Fig. 1.11. In this case we may envisage the kink as due to the 'warping' of the underlying spacetime geometry, rather than to the sudden motion of the central particle. So in a sense, the particle is not really the source of the radiation – that role is reserved for the spacetime curvature. Indeed, radiation of either variety can be generated by 'fixing' the particle (charged and/or massive) and wiggling a massive body around within the electric and/or gravitational field of the fixed particle. The wiggling of the buckled geometry surrounding the massive body shakes off vibrations from the electric and/or gravitational field of the fixed particle, in which the massive body is immersed.

The fact that the generation of gravitational radiation by gravitational forces is so strongly suppressed by the equivalence principle compared with the electromagnetic analogue means that 'gravitonics', unlike electronics, seems to be a very long way off.

### 3.3    Gravity waves from stars

The most obvious astronomical sources of gravity waves are collections of stars orbiting among each other in a complicated way. The motions of these enormous masses set up strong gravitational disturbances which spill out of the system as gravity waves, rather like the turbulent water waves that are produced by power boats milling about in a group.

To estimate the power output we can use the general formula (3.8). The internal power flow will be approximately the internal energy divided by a characteristic time, which could be taken to be a typical orbital period. This would then represent the duration required for the system to change shape appreciably. As remarked in the previous section, because this time is determined by gravitational rather than, say, electromagnetic forces, the gravity wave output is very weak.

If the total mass is $M$, and the radius of the cluster of objects is $R$, a typical orbital period would not differ much from the Kepler value for a body orbiting a point mass $M$, which is $2\pi R^{\frac{3}{2}}(GM)^{\frac{1}{2}}$ (see page 28). Note that this quantity is also comparable to the time taken for a body to fall freely from the periphery to the centre of the system, or alternatively the implosion time if the entire system suffers gravitational collapse. The internal kinetic energy is, on average, (minus) one-half the potential energy, according to the so-called virial theorem (see also page 29). If the mass $M$ is distributed more or less uniformly throughout a spherical volume of radius $R$, then the potential energy of the self-attraction of this sphere of matter is about $-GM^2/R$, so we may take the kinetic energy to be about $GM^2/2R$. The internal power flow is thus

$$\frac{\text{kinetic energy}}{\text{characteristic time}} = \frac{GM^2/2R}{2\pi R^{\frac{3}{2}}(GM)^{\frac{1}{2}}},$$

so the power output of gravitational radiation is about

$$\frac{G^3 M^5/R^5}{c^5/G} = \frac{G^4}{c^5}\left(\frac{M}{R}\right)^5 \tag{3.9}$$

ignoring small factors like $\pi$. The important feature of this result is that it is highly sensitive to the value of $M/R$. As we shall see, the maximum value of this quantity is $c^2/2G$, which occurs when $R = 2GM/c^2$, the gravitational radius, $r_g$ (see equation (3.7)). Substituting this maximum value into (3.9) yields $c^5/G$ (ignoring small numerical factors), which is the 'standard' power output of $3.6 \times 10^{52}\,\mathrm{J\,s^{-1}}$. If $M/R$ becomes larger than this, the system turns into a black hole, and the gravitational radiation is unable to escape. Thus $c^5/G$ represents the maximum power output that is physically attainable.

The simplest composite system is a binary star, which consists of two stars orbiting around their common centre of gravity (see Fig. 3.3). Billions of binary stars exist in our Galaxy. For two stars of equal mass $M$, in circular orbits of radius $R$, the quadrupole moment formula (3.3) yields

$$\frac{64G^4}{5c^5}\left(\frac{M}{R}\right)^5, \tag{3.10}$$

which has the characteristic $(M/R)^5$ of the general expression (3.9).

To get some idea of the numbers involved, suppose two stars of solar mass are in circular orbit $10^7$ km apart. The power output is then about $10^{19}\,\mathrm{J\,s^{-1}}$. Compare this result to the heat output of $10^{26}\,\mathrm{J\,s^{-1}}$ from the Sun, and one sees that the power radiated gravitationally from such a binary system is quite impressive. Nevertheless, if it were situated, say, 1000 pc from Earth (3262 light years, or about $3 \times 10^{16}$ km, typical of stars in our neighbourhood of the Galaxy), the flux of wave energy at the Earth's surface is still minute – about $10^{-21}\,\mathrm{J\,m^{-2}\,s^{-1}}$, many orders of magnitude below detectability.

Formula (3.10) is very sensitive to the orbital radius $R$, so it seems reasonable to ask how low $R$ can be expected to be. Obviously $R$ must be greater than the diameter of the stars concerned, but in practice tidal stresses cause severe disruption to a

Fig. 3.3. Two stars orbiting around their common centre of gravity, $G$, will emit a steady flux of long wavelength gravitational radiation, with frequency = 1/(orbital period).

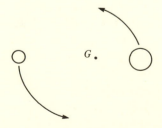

binary star system even for $R$ somewhat greater than this. Typical stars have diameters of the order $10^6$–$10^7$ km, so it is impossible to improve on the above example by more than a few powers of ten.

One of the most promising candidates is the double star system known as $\iota$ Boo, which has a power output of about $10^{23}\,\text{Js}^{-1}$, and being less than 12 pc from Earth it produces a flux of gravitational radiation at the Earth's surface of $10^{-13}\,\text{Jm}^{-2}\text{s}^{-1}$. Table 3.1 lists a number of other binary star systems of interest.

Although individual binaries are unlikely to produce enough radiation to be detectable in the foreseeable future, the cumulative effect of all the binaries in the Galaxy could be considerable. It has been estimated that about $10^{28}\,\text{J}$ escapes from the Galaxy every second as a result of this activity.

Some highly evolved stars are much more compact, however. For example, a white dwarf might be the size of the Earth, while neutron stars or black holes (see below) are typically a few kilometres across. If two such objects of one solar mass orbited 1000 km apart they would complete one orbital period in a mere 0.4 s! The power output from gravitational radiation would be a colossal $3 \times 10^{39}\,\text{Js}^{-1}$. At the Earth's surface, this still only represents a flux of $0.3\,\text{Jm}^{-2}\text{s}^{-1}$ if the system is 1000 pc away, so even in this case the wave intensity is too low for direct detectability with Earth-based equipment.

Even individual neutron stars could be a prolific source of gravity waves. These objects are thought to be remnants of massive stars that have become unstable and exploded. From the theory of stellar evolution it is known that high mass stars burn up their nuclear fuel much more rapidly than the Sun. Most stars consist mainly of hydrogen, which is slowly converted to helium through nuclear reactions which take place in the hot, dense core of the star. In the later stages, the helium can also be used to release nuclear energy, resulting in more complex nuclei being synthesized in the core. These in turn provide more nuclear fuel until heavy elements such as iron are produced. The heavier elements are generally less efficient fuels, and those beyond iron in the periodic table actually provide no net energy gain from the fusion process.

When stars reach this late stage of heavy element fusion, their

Table 3.1. *Gravity waves from binary stars*

| Binary | Period | Mass | Distance from Earth (pc) | $\tau$ (orbital decay time) | $(-dE/dt)_{grav}$ (J s$^{-1}$) | Gravitational radiation at Earth (J m$^{-2}$ s$^{-1}$) |
|---|---|---|---|---|---|---|
| $\eta$ Cas | 480 yr | 0.94 0.58 | 5.9 | $3.8 \times 10^{25}$ yr | $5.6 \times 10^{3}$ | $1.4 \times 10^{-32}$ |
| $\xi$ Boo | 149.95 yr | 0.85 0.75 | 6.7 | $1.5 \times 10^{24}$ yr | $3.6 \times 10^{5}$ | $6.7 \times 10^{-31}$ |
| Sirius | 49.94 yr | 2.28 0.98 | 2.6 | $2.9 \times 10^{22}$ yr | $1.1 \times 10^{8}$ | $1.3 \times 10^{-27}$ |
| Fu 46 | 13.12 yr | 0.31 0.25 | 6.5 | $1.3 \times 10^{22}$ yr | $3.6 \times 10^{7}$ | $7.1 \times 10^{-29}$ |
| $\beta$ Lyr | 12.925 day | 19.48 9.74 | 330 | $2.8 \times 10^{12}$ yr | $5.7 \times 10^{21}$ | $3.8 \times 10^{-18}$ |
| UWCMa | 4.393 day | 40.0 31.0 | 1470 | $3.3 \times 10^{10}$ yr | $4.9 \times 10^{24}$ | $1.9 \times 10^{-16}$ |
| $\beta$ Per | 2.867 day | 4.70 0.94 | 30 | $1.3 \times 10^{12}$ yr | $1.4 \times 10^{21}$ | $1.3 \times 10^{-16}$ |
| WUMa | 0.33 day | 0.76 0.57 | 110 | $2.5 \times 10^{10}$ yr | $4.7 \times 10^{22}$ | $3.2 \times 10^{-16}$ |
| WZSge | 81 min | 0.6 0.03 | 100 | $4.9 \times 10^{6}$ yr | $3.5 \times 10^{22}$ | $2.9 \times 10^{-16}$ |
| 10 000 km binary | 12.2 s | 1.0 1.0 | 1000 | 13.0 yr | $3.25 \times 10^{34}$ | $2.7 \times 10^{-6}$ |
| 1000 km binary | 0.39 s | 1.0 1.0 | 1000 | 11.4 h | $3.25 \times 10^{39}$ | $2.7 \times 10^{-1}$ |

Mass of each component star is shown, in units of one solar mass. The final two entries are hypothetical, very close binaries involving two one-solar-mass objects separated by 10 000 km and 1000 km respectively. Data taken from M. J. Rees, R. Ruffini and J. A. Wheeler, *Black holes, gravitational waves and cosmology* (Gordon and Breach, London, 1974).

energy source starts to fail and they begin to shrink under their own gravity. The highly dense core can, under some circumstances, shrink so fast that it collapses catastrophically in much less than a second. Astronomers believe that the gravitational collapse of the core can produce intense shock waves and bursts of neutrinos – very weakly-interacting subatomic particles. These influences erupt from the imploding core and blast the outer layers of the star apart, releasing more power for a few days than perhaps an entire galaxy. These stellar kamikaze outbursts are known as supernova explosions and occur on average perhaps two or three times a century in each galaxy. The last recorded supernova event in our own Galaxy was in 1604.

The fate of the collapsing core depends critically on its mass. If the core mass exceeds three solar masses it seems an inevitable consequence of Einstein's general theory of relativity that a black hole will result. For cores about one solar mass or less, a neutron star is possible. This is an object in which gravity is so intense that even the atoms collapse under their own weight and are crushed into neutrons. A neutron star might reach a central density of $10^{18}$ kg m$^{-3}$, i.e., nuclear densities, and is rather like a gigantic atomic nucleus. The surface gravity on such a formidable object could be $10^{12}$ times greater than on Earth, so that a thimbleful of matter there weighs the same as $10^{24}$ kg on Earth.

A typical neutron star is found at the centre of the Crab Nebula, about 1700 pc away in the Constellation of Taurus. This is the remains of a star that was observed as a supernova by oriental astronomers in 1054. Today, the exploded outer layers of the star form a ragged nebulous mass, and near the centre is a neutron star with a mass of about 0.8 solar masses, rotating at the extraordinary speed of 30 revolutions per second.

A rapidly rotating star will bulge out at the equator as a result of centrifugal forces. However, if the body remains axially symmetric there is no changing quadrupole moment, and no gravitational radiation is emitted. A slight deviation from axisymmetry might occur if some of the initial turbulence of the collapse phase became 'frozen' into the solid crust of the star. For example, if the Crab neutron star deviated by only one part in a thousand from axisym-

Fig. 3.4. Supernova. (*a*) In 1937 a supernova explosion was observed in the galaxy IC 4182. The supernova was so bright that the galaxy itself does not show up in this brief exposure. (*b*) Fifteen months later it had faded to the edge of visibility. (*c*) By January 1942 the supernova was no longer visible, even using this longer exposure, which reveals details of the surrounding galaxy. (*Hale Observations, California*)

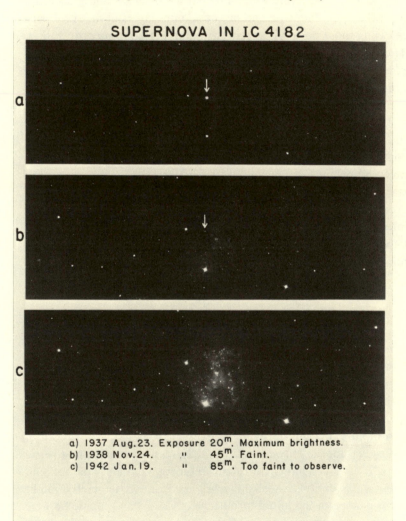

SUPERNOVA IN IC 4182

a) 1937 Aug.23. Exposure $20^m$. Maximum brightness.
b) 1938 Nov.24.    "    $45^m$. Faint.
c) 1942 Jan.19.    "    $85^m$. Too faint to observe.

metry it would acquire a sufficient quadrupole moment oscillation to emit gravitational radiation at the rate of about $10^{31}$ J s$^{-1}$. Although such a rate of emission would deplete the rotational energy of the star by an observable amount, the effect would be complicated by comparable energy losses due to the pulsar mechanism (see section 5.3).

If the neutronic material is stiff enough, then axisymmetry can

Fig. 3.5. Crab Nebula. This ragged cloud of gas is the shattered debris of the supernova explosion observed by oriental astronomers in 1054. At its centre lies a rotating neutron star, advertising its presence by extremely rapid radio pulses. Great bursts of gravitational radiation would have accompanied the supernova violence that formed the neutron star, but the rhythmic gravity wave emission from the continuing rotation of the star is far too weak to detect. (*Lick Observatory, California*)

suddenly fail at a critical speed, and the star's shape may become very ellipsoidal at high angular momentum, leading to prolific gravitational radiation. Unfortunately, too little is known about the properties of ultradense matter for any reasonable assessment of the likelihood of this possibility.

More promising than the rotation of the neutron star is the violence that accompanies its birth, buried inside the turmoil of a supernova. During the implosive–explosive turbulence, the nascent ball of neutrons will be endowed with vibrational, as well as rotational, energy. Crudely speaking, the star will ring. The lowest mode of vibration will be the symmetric breathing mode – a purely radial inflation and deflation with a typical frequency of a thousand cycles per second. There is no quadrupole moment associated with this motion, so no gravitational radiation ensues. High modes of vibration will, however, radiate intensely. Moreover, if the star is also rapidly rotating, then the radial mode is perturbed and can feed energy into the quadrupole modes.

The Crab neutron star has about $10^{42}$ J of rotational energy today, though some has been lost since its formation. One might expect about $10^{43}$–$10^{44}$ J of vibrational energy to accompany the birth of such an object. This energy will be radiated away gravitationally in a time that might be anywhere from 0.1 s to a few days, depending on details.

It is conceivable that this type of process can generate as much as $10^{44}$ J s$^{-1}$, and if located only 100 pc away it would yield a flux of $10^{6}$ J m$^{-2}$ s$^{-1}$ at Earth. This is within the range of current detectors, if they are tuned in to the frequency of the source vibrations.

Although our knowledge of supernova explosions and neutron star formation is still rudimentary, a plausible scenario might be the following. The core of a massive, old star runs out of nuclear fuel and starts to shrink under its own weight. To conserve angular momentum it begins to rotate more rapidly until, at the point when it collapses to a neutron star, it is considerably flattened. The additional implosion to neutron star dimensions drastically increases the rotation rate to many times per second, and the ball of neutrons becomes flattened almost to a pancake. If the strain proves too much it fragments into a composite system of smaller neutron stars. (The outer envelopes of the original star have by now been blown off.) The loss of energy by gravitational radiation in this turbulent maelstrom is so great that before long the fragments begin to coalesce, each encounter causing an intense burst of waves. Eventually a single system – either a neutron star or black hole – forms, in a highly turbulent condition, and its non-radial 'ringing' motions are slowly damped by the emission of yet more gravitational radiation.

The gravity wave output would therefore consist of a sequence of intense bursts of radiation during the collapse of the star core and the subsequent reorganization of the fragments, followed by a steady and continuous emission due to the ringing motions of the end product. It has been estimated that about 1 per cent of the core rest mass might be radiated in the initial bursts, and perhaps 5 per cent in the ringing. The star may therefore emit more energy in a few years as gravitational radiation than as heat and light during its entire multi-billion-year lifetime.

Even after the neutron star has settled down, it may continue to radiate gravitational bursts as a result of 'starquakes' in its interior. Sudden changes in the radio signals from pulsars have been interpreted as neutron starquakes. The strain energy released during such a quake has been estimated at up to $10^{38}$ J, and a high fraction of this energy could be converted to gravity waves.

Similar violent events in other stars will also lead to gravitational radiation. For example, the stellar explosions known as novae, which are less violent but rather more frequent than supernovae, might radiate $10^{28} \mathrm{Js}^{-1}$. As there are so many of these events, there is a good chance that such an outburst will occur only a few hundred parsecs away, so that although the energy output is relatively modest, the flux at Earth may be, say, $10^{-11} \mathrm{Jm}^{-2}\mathrm{s}^{-1}$. However, this is still many orders of magnitude too low for detectability.

It is interesting to compare the gravitational radiation power output due to the gross motion of a star with that due to the thermal motions of its constituent atoms. In the core of the Sun, the temperature is about $10^7$ K, so the thermal energy per particle is around $10^{-16}$ J. The core is so dense that the electrons (which undergo the most vigorous activity) are subject to quantum forces, so we take as the characteristic time $\hbar/$(energy of electron) $\sim 10^{-18}$ s. The power flow per electron is thus $\sim 10^2 \mathrm{J s}^{-1}$, giving a gravity wave power rate of $3 \times 10^{-49} \mathrm{J s}^{-1}$. The solar core contains about $5 \times 10^{56}$ electrons, yielding a total power rate of $\sim 10^8 \mathrm{J s}^{-1}$, or a hundred megawatts. This is to be compared with $10^{26} \mathrm{J s}^{-1}$ radiated as heat and light, and $10^{-4}$ Js emitted as gravity waves by virtue of the Earth's motion around the Sun (the latter being not enough to power a single light bulb!).

## 3.4    Gravity waves from black holes

Undoubtedly the most prolific sources of gravitational radiation are the so-called black holes. Although the emission of gravity waves by other systems is certainly important for their structure and development, when it comes to detecting gravity waves on Earth, hopes are mainly pinned on the sort of extreme violence which can only accompany the formation and subsequent catastrophic experiences of black holes.

In section 3.1 it was pointed out that a power output approaching the limiting case of $c^5/G$ can only occur in systems close to their gravitational radii $(r_g = 2GM/c^2)$, moving at relativistic speeds $(v \approx c)$.

When a star, or other object, is compressed to near this radius, the curvature of spacetime $(\sim GM/c^2 r^3)$ becomes comparable to the curvature of the star's surface $(1/r_g{}^2 \sim c^4/G^2 M^2)$. This indicates that severe distortions of space and time will occur near the surface. The Newtonian expression for the gravitational energy of a particle of mass $m$, $-GMm/r$, valid at large $r$, becomes modified near $r_g$ by relativistic effects to

$$mc^2[(1 - 2GM/rc^2)^{\frac{1}{2}} - 1],$$

which approaches $-mc^2$ (i.e., the entire mass–energy of the particle) as $r$ approaches $r_g$. Once the star's radius drops below this value, a particle at the surface can never escape. Even light becomes trapped by the curvature of spacetime.

It can be shown that no material can withstand the enormous gravity at the gravitational radius. The star must implode, and calculations show that it collapses out of existence in perhaps a microsecond of star-time. There is no agreement on what happens to the imploded star, but some mathematicians conjecture that it will tear spacetime apart and disappear from the Universe altogether at a sort of edge of spacetime, where gravity is so intense that the curvature rises without limit. Be this as it may, the region of space once occupied by the star is left empty and, because no light can escape, it is invisible to the exterior Universe. Therefore the region of space which the star once filled is now left as a black hole.

Although nothing, including gravity waves, can escape from actually inside the hole, the severe disruption of the surrounding spacetime caused by the exceedingly rapid implosion of the star will cause a tremendous release of gravitational radiation energy. Exactly spherical collapse will not produce gravity waves because of the absence of a changing quadrupole moment, but in practice rotation and turbulence would feed a high fraction of the available energy into asymmetric motions.

We may make a crude estimate of the energy output from the collapse of an object to form a black hole using the very general order-of-magnitude formula (3.9). The output will occur in a short burst as the system collapses. The collapse time is about $R^{\frac{3}{2}}/(GM)^{\frac{1}{2}}$ (see page 67) so the total energy output will be

$$E = \frac{R^{\frac{3}{2}}}{(GM)^{\frac{1}{2}}} \times \frac{G^4}{c^5} \left(\frac{M}{R}\right)^5.$$

For a black hole $R \sim GM/c^2$, so this expression reduces to simply

$$E = Mc^2,$$

i.e., the entire mass–energy of the system is radiated away as gravity waves. Of course, this result is an upper limit, and the exact fraction of $Mc^2$ will depend critically on the details of the particular collapse. In most cases it probably lies between 1 per cent and 10 per cent of $Mc^2$ (though some detailed calculations suggest a range of values somewhat less than this). This is an impressive quantity of energy (up to about $10^{46}$ J for a solar-mass star). Moreover, black holes could also form from much larger objects than individual stars, so even a modest conversion efficiency into gravity waves could mean an enormous output.

For example, a large number of stars in a dense cluster might collapse together to form a superhole. Our Galaxy contains about 300 globular clusters, consisting typically of 100 000 stars, formed during the very early life of the Milky Way. If, say, one thousand of these stars began to congregate near the centre of the cluster and spiral together, they might approach their collective gravitational

Fig. 3.6. Globular cluster NGC 288. During the formation of our Galaxy, enormous clusters of stars separated out from the primeval gases in spherical blobs measuring dozens of light years in diameter. The high concentration of stars near the centre suggests that massive black holes lurk there. If the holes swallow stars, or even collide with each other, enormously powerful bursts of gravitational radiation will be emitted. (*Anglo-Australian Observatory, New South Wales*)

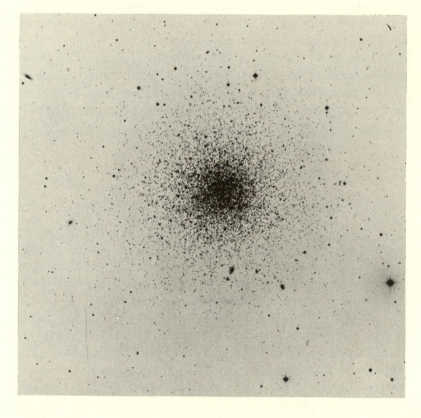

radius and fall into each other to form a huge black hole, which would radiate thousands of times more energy than individual stellar collapse. There is the inevitable trade-off that the former events would be much more rare than the latter, so that one has to search a wider volume of space. Taking into account possible collisions between superholes as well as their formation, the

Fig. 3.7. Quasar 3C 273. In the early 1960s astronomers began discovering 'quasi-stellar' objects, or quasars. Their starlike appearance belies their great remoteness; the above example is relatively close to Earth at about half a billion parsecs. Quasars are believed to be exceedingly dense and very energetic. The extremely violent processes that must occur in their interiors, some evidence of which can be deduced from the faint jet of material being ejected from 3C 273, are ideal for the generation of gravity waves. Many astronomers believe quasars contain black holes. (*Kitt Peak National Observatory, Arizona*)

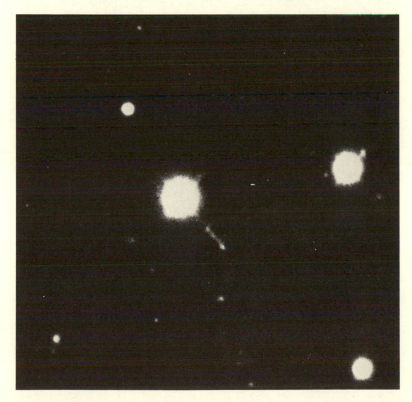

American astronomer Kip Thorne has estimated one event per month in a volume of space of $10^9$ pc radius – which is nearly the entire observable Universe. If a high fraction of the imploding matter is converted to gravity waves, then even from such enormous distances these bursts should produce a flux at Earth in excess of any of the less violent events within our Galaxy.

Still more massive black holes could form in the centres of certain active galaxies, or in quasars – highly compact, superenergetic objects on the edge of the observable Universe. The superholes here could contain as much mass as a billion stars. These events would be so rare that they may only occur once every hundred years or more in the entire observable Universe. Nevertheless, they would produce pulses of gravity waves at Earth many millions of times stronger than black hole events within our own Galaxy.

Some astronomers believe that galaxies contain halos of supermassive black holes formed during the first 100 million years of the Universe, following the big bang, when the cosmological material was packed much more densely. Indeed, an appreciable fraction of the mass of the Universe might be locked up in this form. If this were correct, the number of violent events accompanying the birth of these objects would have been so great that the Universe should now be continually bathed in their gravity waves with an intensity at Earth which might conceivably be millions of times that due to processes involving black holes of solar mass within our Galaxy.

Although the formation of a black hole is the most hopeful source of detectable gravity waves, once formed the hole can still produce radiation if it swallows other matter. Even whole stars could be gobbled up this way, releasing a large proportion of their mass as gravity waves.

Detailed calculations have been performed to determine the efficiency of conversion of mass–energy into gravitational radiation when a small body of mass $m$ falls into a large black hole of mass $M$ (see Fig. 3.8). To obtain a rough idea of the output we can use the general expression (3.8). As remarked on page 76, when a particle of mass $m$ is lowered to the gravitational radius of another body, its gravitational binding energy is equal to $mc^2$ minus its entire rest mass–energy. Consequently, if the particle is dropped into the hole, its kinetic energy will rise to a high fraction of its rest mass, say $\frac{1}{2}mc^2$ on average. This is so high that the particle travels at near the speed of light, $c$. Most of the gravity waves are emitted from the strong field region close to $r_g$. However, once inside $r_g$, the particle has entered the black hole and no gravitational radiation can escape. Therefore, most of the energy emanates from that portion of the trajectory lying between, say, $5r_g$ and $r_g$. Ignoring acceleration, the particle takes roughly $4r_g/c$ to fall this distance. The power flow is thus

$$\frac{\text{kinetic energy}}{\text{characteristic time}} \sim \frac{\frac{1}{2}mc^2}{4r_g/c} = \frac{mc^5}{16GM},$$

so from (3.8) the power output in gravitational radiation will be about

$$\frac{G}{c^5}\left(\frac{mc^5}{16GM}\right)^2 \approx 0.004\,\frac{c^5}{G}\left(\frac{m}{M}\right)^2.$$

The total radiated energy will be roughly this quantity multiplied

Fig. 3.8. When a small object falls into a large black hole it emits gravity waves strongly in the region $r =$ several $r_g$. As $r_g$ is approached, gravitational 'redshift' (see page 91) severely reduces the frequency and energy of the waves, cutting off the output completely as the object crosses into the hole, and oblivion.

$r = r_g$

by the characteristic free-fall time, or about

$$0.03 \left( \frac{m}{M} \right) mc^2. \tag{3.11}$$

The coefficient 0.03 is typical of the sort of values obtained from detailed calculations using idealized models; these values range from 0.01 to about 1 depending on the details, such as angle of infall, rotation of the hole, etc. Note that if $m \ll M$, then the output is only a very small fraction of the rest mass of the sacrificed particle.

However, when $m \sim M$, the output could be considerable. This corresponds to the case of, say, a solar-mass black hole swallowing a neutron star, or the collision of two equal-mass black holes. Very extensive and ambitious computer calculations have been performed by the American astrophysicist Larry Smarr to determine the energy output from the collision of two non-rotating black holes. The results seem to indicate an efficiency coefficient somewhat below 0.01, and even as low as 0.001. Nevertheless, if the holes were rotating one could still expect a high fraction of the total mass to be radiated. This is an extraordinary thought. A star will typically spend many billions of years radiating heat and light as a result of which about 1 per cent of its mass–energy is lost. If it then collapses to a black hole which collides with another it may loose, say, ten times this quantity ($\sim 10^{46}$ J) in a single burst of gravity waves. The flux of energy at the Earth's surface, if such an event occurred at the centre of our Galaxy, would be around $10^4 \, \mathrm{J \, m^{-2} \, s^{-1}}$.

It is interesting to ask whether there is an upper limit to the amount of energy that can be radiated by two colliding black holes. Professor Stephen Hawking of the University of Cambridge has proved a fundamental theorem that in any process the total area of black holes cannot decrease, and this does indeed set a maximum efficiency factor for energy conversion. (This is a direct analogy with thermodynamics, where the second law, which forbids entropy to decrease, limits the energy–heat conversion efficiency of heat engines.)

The area of a non-rotating (and uncharged) hole is

$$4\pi r_g{}^2 = \frac{16\pi G^2 M^2}{c^4},$$

i.e., proportional to $M^2$. Energy, however, is proportional to $M$ (i.e., $Mc^2$). If two equal-mass (hence equal-area) holes coalesce, the final area of the single coalesced hole must be at least as great as the total area of both original holes, i.e., $8\pi r_g^2$. The mass of the final single hole that yields this area is $\sqrt{2}M$, so we started with a total mass $2M$ and ended with $\sqrt{2}M$. Thus $(2-\sqrt{2})M$, or about 29 per cent of the total original mass $2M$, has been lost. This 29 per cent is the theoretical upper limit on the conversion efficiency for black hole mass into gravitational radiation.

If the holes are rotating, the upper limit is raised to 50 per cent, and if they also carry the maximum permissible quantity of electric charge, it can reach as high as 65 per cent. In realistic astrophysical situations, however, conversion efficiencies would probably be at least an order of magnitude lower than these limiting values.

### 3.5    Cosmological sources

Most astronomers today believe that the Universe began with a big bang about 15 billion years ago. In the 1920s Edwin Hubble discovered that the distant galaxies are receding from us and each other in a state that is best described as an expansion of the entire Universe. Modern optical and radio telescopes indicate that this expansion is highly uniform in all directions.

If the galaxies are receding now, they must have been closer together in the past. Extrapolating backwards in time they would, at their present rate of recession, have been compressed completely together several billion years ago. However, to escape their mutual gravity, the galaxies must have been expanding much faster then than now, and taking this into account, one arrives at a time about 15 billion years ago when all the matter in the Universe exploded from a hot, dense, firey maelstrom.

General relativity provides a natural explanation for this expansion as an inflation of space. The 'elastic' spacetime of Einstein can be envisaged as stretching in all directions, sweeping the galaxies along with it. Frequently analogy is made with a rubber sheet covered in dots to represent the galaxies. As the sheet is stretched so each dot moves away from every other (see Fig. 3.9). Each dot regards itself as at the centre of a uniform pattern of expansion.

Note, however, that the dots are not moving towards, or away from, any particular place. The big bang was *not*, therefore, the explosion of a lump of matter at one place into a pre-existing void. It was the eruption of space, with matter 'embedded' in it, in an explosive expansion motion. Many physicists believe that the big bang actually represents the *creation* of spacetime and matter, so that there was simply nothing at all – no places, no things, no moments – before that event.

Be that as it may, one can study in detail the epochs of intense and violent activity that followed the initial eruption. Of special interest are the turbulent and chaotic motions of matter and spacetime that would have generated enormous quantities of gravitational radiation. The problem about estimating the amount of gravity waves is that the epochs of greatest violence were also the earliest, about which almost nothing is known.

It was mentioned above that the present expansion of the Universe is remarkably uniform. This could be interpreted in two ways. It could be that the Universe was created in a chaotic state, and that the primeval turbulence has been damped out, with a significant fraction going into gravitational waves. From this viewpoint, the Universe should be filled with gravitational radiation of great energy. On the other hand, the present smoothness and uniformity might indicate that the Universe began smoothly with very little turbulence to generate gravity waves. (The uniform expansion motion itself does not produce gravity waves as it is too

Fig. 3.9. Expansion of the universe. Modelled by a rubber sheet, one can see how, as the sheet is stretched, the gaps between all the dots grow. If the sheet is infinite (or closed into a sphere) it has no centre, yet every dot (e.g., *A*) sees all neighbouring dots (e.g., *B,C*) steadily receding from it.

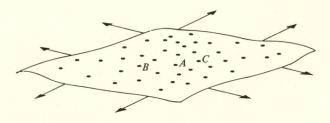

symmetrical; there is no changing quadrupole moment.) The confirmation as to which view is correct will have to await better knowledge of the overall and relative efficiencies of various competing damping mechanisms.

In spite of these problems, there are fundamental physical reasons why at least *some* turbulence and clumpiness must have accompanied the birth of the Universe. First, the galaxies exist. If we envisage the primeval fire as a smoothed-out fluid of mainly hydrogen gas, then if, by random fluctuation, the gas begins to accumulate in a certain region, it will form a nucleus of matter at slightly enhanced density which, by its own gravity, will begin to attract other matter. The density perturbation will thus grow, and might eventually scoop up enough of the surrounding material to form a galaxy.

However, this aggregation of matter has to compete with the universal expansion, which was very much more rapid then than now. Calculations show that the growth rate of these purely chance accumulations is insufficient to produce the galaxies that we observe today. The conclusion seems inevitable that the primeval cosmos must have already contained at the outset a fair amount of clumpiness to act as gravitating centres for the protogalaxies. If this was so, then, as pointed out by the British astronomer Martin Rees, gravity waves would have been produced when these huge clumps were banging around among the primeval gases.

To obtain a very crude estimate of the gravitational radiation output we can use (3.8) once again. A typical galaxy has a mass of around $10^{41}$ kg, and if a sizeable fraction of this were contained in a protogalaxy moving at, say, 10 per cent of the speed of light, the kinetic energy would be around $10^{56}$ J. The present density of galaxies is about one per $10^{69}$ m$^3$, but at the time of galaxy formation, the universe was only about one-tenth of its present age and the density of protogalaxies was between 30 and 100 times greater – approximately one galaxy per $10^{67}$ m$^3$, corresponding to an average separation of around $10^{22}$ m. The time taken for a protogalaxy to 'bounce' between two others at $0.1c$ would thus be about $10^{22}/(3 \times 10^7) \approx 3 \times 10^{14}$ s, i.e., a few million years. However, most of the gravity wave output would occur during the collisions

Fig. 3.10. Field of galaxies. A cluster of distant galaxies is shown, many of whose members appear simply as fuzzy blobs. In an earlier phase of the Universe, when these objects were still forming from the primeval gases amid much turbulence, intergalactic collisions would have generated vast quantities of very long wavelength gravitational radiation, which still bathes the Universe today. (*Anglo-Australian Obersvatory, New South Wales*)

themselves, which would have occupied somewhat less time than this – say $10^{13}$ s. Taking the latter as our characteristic time, the power flow would be

$$\frac{\text{kinetic energy}}{\text{time}} \sim \frac{10^{56}}{10^{13}} = 10^{43} \, \text{J} \, \text{s}^{-1},$$

yielding a gravity wave power output of about

$$\frac{10^{86}}{c^5/G} \approx 3 \times 10^{33} \, \text{J} \, \text{s}^{-1}.$$

If this activity continued for a billion or more years it would generate a total energy output per galaxy of, say, $10^{50}$ J, which corresponds to an energy density throughout the Universe of around $10^{50}/10^{67} = 10^{-17} \, \text{J} \, \text{m}^{-3}$. However, the subsequent expansion of the Universe will have reduced this quantity by a factor of several hundred, and so we arrive at a present background energy density of gravity waves from the epoch of galaxy formation to be around $10^{-19} \, \text{J} \, \text{m}^{-3}$. To get some feeling for this quantity, we can convert it into a mass density using $E = mc^2$, and arrive at $10^{-36} \, \text{kg} \, \text{m}^{-3}$. This should be compared with the density of galactic matter in the Universe, which is about $10^{-28} \, \text{kg} \, \text{m}^{-3}$, or a hundred million times greater. Later we shall discover whether or not this quantity of gravitational radiation could be significant.

A second fundamental reason for expecting some primeval turbulence concerns the quantum theory of gravity (see section 5.4). According to the Heisenberg uncertainty principle, the energy of a system is likely to fluctuate by an amount $\Delta E$ during a period of time $\Delta t$, where $\Delta E$ and $\Delta t$ are related by

$$\Delta E \, \Delta t \approx \hbar \tag{3.12}$$

$\hbar$ being Planck's constant divided by $2\pi$. If this principle is applied to the gravitational field, one can envisage sudden and unpredictable changes in field energy, which in turn represent fluctuations in spacetime curvature, or tidal gravity.

To estimate the scale of these fluctuations we may consider that the ripples of geometry produced by the quantum effects propagate like gravitational waves at the speed of light. In a sphere of radius $r$, therefore, the time taken for the disturbance to reach the surface of the sphere is $\Delta t = r/c$, so from (3.12) we obtain an energy fluctuation of about $\Delta E \approx hc/r$. These fluctuations will ordinarily be minute. In an atomic-sized sphere, for example, $\Delta E \approx 10^{-16}$ J, which has a mass of only $10^{-33}$ kg. (Compare this with the mass of an atom of hydrogen, $\approx 10^{-27}$ kg.) The gravitational effect of such a tiny mass is utterly negligible.

We may estimate the importance of this quantum gravity by calculating the gravitational self-energy of the fluctuation, i.e., the energy required to pull $\Delta E$ apart against its own gravity. Newtonian gravitational theory yields $-GM^2/r$ for the self-energy of a spherical mass $M$, so the self-energy due to Heisenberg quantum fluctuations is about

$$-\frac{G(\Delta E)^2}{rc^4} \approx -\frac{G\hbar^2}{r^3c^2}.$$

In an atomic-sized volume there is a gravitational self-energy of only about $-10^{-66}$ J, which is fifty powers of ten less than the fluctuation energy itself.

If we choose the length scale $r$ small enough, however, a point is reached where the quantum energy will, when it fluctuates into existence, bind itself strongly together under its own gravity, perhaps even becoming a temporary quantum black hole. When the gravitational binding self-energy reaches a value comparable to the energy $\Delta E$ itself, therefore, drastic disturbances will occur in the gravitational field and geometry of spacetime.

This disruptive regime is approached when

$$\frac{G(\Delta E)^2}{rc^4} \approx \Delta E \approx \frac{\hbar c}{r},$$

or when $\Delta E \approx (\hbar c^5/G)^{\frac{1}{2}}$. This corresponds to a mass of $(\hbar c/G)^{\frac{1}{2}} \approx 2 \times 10^{-8}$ kg. The length scale $r$ is then given by $r \approx \hbar c/\Delta E \approx (\hbar G/c^3)^{\frac{1}{2}} \approx 2 \times 10^{-35}$ m, and the characteristic fluctuation time, $r/c \approx (\hbar G/c^5)^{\frac{1}{2}} \approx 5 \times 10^{-44}$ s. These distance and time scales are some twenty powers of ten smaller than nuclear values, and utterly beyond the range of direct measurability.

The length scale $(\hbar G/c^3)^{\frac{1}{2}}$ is known as the Planck length, because Max Planck first pointed out that a new fundamental length scale could be constructed from the constants $\hbar$, $c$ and $G$. At Planck scales, it would be expected that quantum gravity effects would be of great importance. We shall return to this topic in section 5.4. For now, our interest centres on the fact that, whatever happened afterwards, back at $5 \times 10^{-44}$ s (called the Planck era) after the initiation of the big bang, the Universe cannot possibly have been smooth, but must have witnessed wild and turbulent quantum fluctuations.

Picturesquely, this primeval chaos can be envisaged by regarding space as a violently quivering jelly. As the universe expanded, the quivers would have died away, but some relic of the shakes must remain with us today in the form of gravity waves, bathing the entire Universe.

The energy density of primeval gravitational radiation can, as usual, be estimated from (3.8). At the Planck era, the density of quantum energy in the Universe is given by (Planck energy)/(Planck volume). The characteristic fluctuation time is the Planck time, so the power flow per unit volume is

(Planck energy)/(Planck volume × Planck time).

To estimate the total gravity wave energy we may assume that peak power output only lasted a few Planck times; after this, the expansion of the Universe reduced the relative importance of the fluctuations. We therefore arrive at a total energy density of background primeval gravitational waves:

$$\frac{G(\text{Planck energy})^2}{c^5(\text{Planck volume}) \times (\text{Planck time})}$$

$$= \frac{\dfrac{G}{c^5} \times \dfrac{\hbar c^5}{G}}{\left(\dfrac{\hbar G}{c^3}\right)^{\frac{3}{2}} \times \left(\dfrac{\hbar G}{c^5}\right)^{\frac{1}{2}}}$$

$$= \frac{c^7}{\hbar G^2} \approx 10^{114} \, \text{J m}^{-3},$$

which is the Planck energy density (equivalent to $10^{97}$ kg m$^{-3}$).

Since the Universe was then very hot, the gravitational radiation would probably have been in thermal equilibrium at a temperature corresponding to the above energy density. Using Stefan's law, which relates temperature to energy density, we can deduce that the temperature at the end of the Planck era was $10^{32}$ K (compare $10^7$ K at the centre of the Sun).

As the Universe expanded, the temperature fell. The reason for this is that the gravity waves are embedded in an expanding space, so as the space stretches, so the wavelength grows in proportion. Because temperature is inversely proportional to wavelength, the temperature falls in inverse proportion to the stretching length scale. Einstein's equations predict that the length scale grows like (time)$^{\frac{1}{2}}$ so long as radiation dominates over matter, which remained the case up to about $10^5$ years. After this the radiation was too cool to have much gravitational effect, and the length scale grew like (time)$^{\frac{2}{3}}$. Thus, the temperature at $10^5$ years was

$$10^{32} \times \left(\frac{\text{Planck time}}{10^5 \text{ years}}\right)^{\frac{1}{2}} \approx 10^4 \text{K},$$

and since then it has cooled by a further factor

$$\left(\frac{10^5 \text{ years}}{10^{10} \text{ years}}\right)^{\frac{2}{3}} = 10^{-10/3},$$

giving a present (i.e., at $10^{10}$ years) background temperature of a few K. This gravitational radiation is accompanied by primeval heat radiation. The latter has been detected, and has a temperature of about 3 K, which corresponds to an energy density of about $10^{-14}$ J m$^{-3}$ or $10^{-31}$ kg m$^{-3}$.

### 3.6    Characteristics of the radiation

Before leaving the subject of sources, something should be said about the characteristics of the gravitational radiation likely to emanate from astrophysical and cosmological sources.

Gravitational radiation from space is of two types. First there are short, sharp bursts which come from sudden and catastrophic events such as supernova explosions or the collision of two black holes. Then there are continuous waves due to the thermal background from the big bang, from binary stars and pulsars, and from the integrated effect of many distant catastrophic events.

The wavelength and frequency of the waves are connected by the formula

$$\text{wavelength } \lambda = \frac{\text{speed}}{\text{frequency}} = \frac{2\pi c}{\omega}.$$

These quantities will vary all the way from short waves in the thermal background ($\lambda \sim 10^{-3}$ m) up to waves with a length comparable to the Hubble radius ($\lambda \sim 10^{26}$ m, the size of the Universe) from large scale cosmological disturbances such as galaxy formation. In between will be the waves corresponding to astrophysical processes such as black hole events.

In general, for a catastrophic event, the pulse of radiation will contain many frequencies. The spread in frequency will be comparable to (duration of the pulse)$^{-1} \sim$ (free-fall time in source)$^{-1}$. It is clear that the higher frequencies come from the later stages of the collapse, because the system gathers speed as it implodes. However, when the gravitational radius is approached another effect becomes important. As the gravity waves leave the star, they must climb up against the escalating surface gravity. Gravity waves, being a form of energy, are subject to this background gravity, and will lose energy as they climb (see, for example, Fig. 3.8). Physically this is manifested as a loss of frequency, an effect called the *gravitational redshift* when it happens to light (red being the long wavelength of optical light). Therefore, the frequency which reaches the Earth is somewhat lower than that which the system originally generated.

When the gravitational radius $r_g$ itself is reached, the frequency shift becomes unlimited, i.e., all frequencies are shifted to zero, which means that no gravity waves at all can escape. This corresponds to the formation of a black hole. There will thus be no frequencies *above* $c/r_g$, since these waves are likely only to be produced inside the hole, when the star's radius is less than $r_g$. A pulse of radiation resulting from gravitational collapse to a black hole is therefore likely to extend from zero up to $c/r_g$ and then to cut off sharply. For a ten solar-mass black hole, this top frequency is around $10^3$ Hz.

In contrast to the catastrophic and turbulent events, gravity waves from continuous sources such as binary stars and neutron star vibrations will have rather narrow frequency spectra. Thus, a study of the frequency distribution of gravity waves can reveal details about the internal structure of the source. If in addition the polarization characteristics of the waves can be studied, still more information about the generation mechanism could be provided. In the next chapter, some of the current and planned technology to accomplish these goals will be described.

# 4    Gravity wave detectors

In the previous chapter some of the natural sources of gravity waves were described. These sources are many and varied, so that the detection of gravity waves promises to open up a rich new field of astronomy.

Gravity waves apparently emanate from the earliest imaginable moments after the creation, from the vicinity of black holes, from the interiors of quasars and neutron stars, from the turbulent depths of star clusters and many other systems. An analysis of this radiation would provide information of incomparable value about these highly inaccessible and remote locations of the cosmos, and would tell us something about the behaviour of spacetime and matter under the most extreme conditions.

The detection of gravity waves would also provide a beautiful check on Einstein's general theory of relativity – the first qualitatively new test of general relativity for over sixty years. In this chapter the nascent technology of gravity telescopes will be described, a technology that could revolutionize astronomy in the decades to come.

## 4.1    Basic principle of the detector

Antennae built to receive electromagnetic waves adorn most domestic rooftops these days. The principle of detecting gravity waves with an antenna is simplicity itself. When gravity waves wash through matter, they set it into vibration. All the experimenter has to do is to look for otherwise unaccountable shakes in his equipment.

Why do gravity waves shake matter? The waves are really ripples of geometry, alternately trying to stretch and shrink distances in the patterns shown in Fig. 2.15 and discussed in the accompanying paragraphs. If the detector is a lump of solid matter, different parts of the body try to fall in different ways in the tides of gravity, so the body is stretched and squeezed, and stresses are set up. However, we do not need to visualize the process like this. Instead, we can regard the vibrations of the detector as due to the conversion of gravity waves into sound waves inside the material – some of the energy from the vibrating spacetime is absorbed by the metal and turned into energy of vibrating matter. In quantum language, gravitons are annihilated and phonons are created.

The efficiency of such an antenna is determined by the fraction of incoming energy flux that can be absorbed and converted into sound (the graviton–phonon coupling strength). The design might be no more complicated than a chunk of aluminium, or two balls joined by a spring, or two pendula suspended a distance apart – anything that will register a vibration of distance as the waves sweep by.

In a radio receiver, there is an electric oscillator which is 'tuned' to the required frequency of the incoming signal. The radio waves excite (electrically) a resonant electric vibration in this internal circuit, which is then amplified and analysed. When searching for gravity waves from a natural source, two strategies may be adopted. One is to build a detector tuned to some expected frequency (e.g., neutron star pulsation rate, binary star rotation period), so the antenna is then most responsive to the signals which one expects to be present. The other approach is to build a broad band detector which will respond to a wide range of incoming signals, on the basis that the strongest waves are likely to come as short bursts from black hole events, the peak frequency of which can only be estimated very crudely.

Before embarking upon these technical features, it is helpful to have some idea of the strength of the effect that we are up against. In chapter 3 the intensities of gravity wave fluxes from various sources were discussed. Many sources of gravitational radiation produce, at least for a short duration, a flux of energy at Earth of between $10^{-10} \mathrm{J m^{-2} s^{-1}}$ and $10 \mathrm{J m^{-2} s^{-1}}$. The highest energy at Earth that could be expected in a reasonable search time (say one year) would be a burst from the collision of two superholes in a very distant galaxy or a quasar, which might produce a total energy of $10 \mathrm{J m^{-2}}$. How vigorously would a detector vibrate in response to a signal of this strength?

A direct way to answer this question is to apply the standard theory of the driven oscillator (see Appendix). However, the same result may be deduced using an argument of great generality from thermodynamics, called detailed balancing. This treatment is, moreover, very easy to follow.

Suppose we have a detector that vibrates naturally at a particular angular frequency $\omega_0$. For concreteness, think of the idealized case of two equal masses $M$ joined by a spring of length $L$ (see Fig. 2.14). If this antenna is immersed in a thermal bath of gravitational waves at temperature $T$, coming from all directions uniformly, then the detector will start to absorb energy from the waves, and vibrate. As at vibrates, so it will begin to *emit* (i.e., reradiate) gravity waves, because its quadrupole moment fluctuates. So long as the rate of emission is below the rate of absorption, the net energy of the detector will rise. Eventually a point must be reached where the vibrations become so vigorous that the detector is in a steady state, with the rate of loss of energy through gravity wave emission just balanced by the rate of energy gain from gravity wave absorption.

Not all the energy absorbed from the waves will go into coherent oscillatory motion of the detector. Some vibrational energy will be converted to heat inside the spring because of friction and other dissipative effects. However, this heat energy will cause the molecules of the spring to jiggle more vigorously, and, on a microscopic scale, this motion will itself generate gravity wave energy. When thermodynamic equilibrium is achieved, the output from these molecular fluctuations will exactly balance the conversion of incoming gravity waves into heat. Thus, the equilibrium motion of the oscillator will be reduced to a random jiggling, and the detector will be at the same temperature as the surrounding bath of gravity waves.

This situation is similar to the electromagnetic case. When a body reaches thermodynamic equilibrium with a surrounding radiant heat bath, the molecular motions cause electromagnetic heat emission at just the same rate as heat energy is absorbed from the surroundings.

We may use this equilibrium to compute the *cross-section*, denoted $\sigma$, for gravity wave absorption, which can be envisaged physically as the effective area presented by the detector to the waves. In order to balance the energy we demand

$$\begin{pmatrix} \text{incoming flux of} \\ \text{radiation} \end{pmatrix} \times \text{cross-section} = \begin{pmatrix} \text{outgoing rate} \\ \text{of radiation} \end{pmatrix}.$$

(4.1)

However, $\sigma$ will be a function of frequency, $\sigma(\omega)$, because the detector will respond more readily to incoming waves with frequency close to the natural frequency of the detector, $\omega_0$. So (4.1) should really be written

$$\int \sigma(\omega) \times (\text{flux per Hz}/2\pi) \, d\omega = \begin{pmatrix} \text{outgoing rate of} \\ \text{radiation} \end{pmatrix}.$$

(4.2)

where angular frequency $\omega$ is measured in units of Hz/$2\pi$ (radians per second).

The variation of cross-section with frequency is shown in Fig. 4.1. The response of the detector is sharply peaked around the natural frequency $\omega = \omega_0$, and falls rapidly to zero outside the

Fig. 4.1. Resonance. Most detectors respond to a narrow range of frequencies around the resonant (natural) frequency $\omega_0$. The width of the range ($\Delta\omega$) is determined by the damping time $\tau$ ($\Delta\omega \approx \tau^{-1}$).

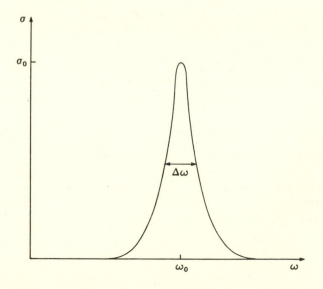

range $\omega_0 \pm \Delta\omega/2$. This general shape is characteristic of driven oscillators, and the very selective frequency response is called *resonance*. The exact shape of the curve is discussed in the Appendix, but here we need only note that the width of the peak is determined by the *damping* time of the oscillator.

To understand this relation, note that when an oscillator is struck and left to vibrate freely it will not continue oscillating for ever, but will slowly decay as damping forces such as friction gradually sap the energy and convert it into heat. The effect of damping is shown in Fig. 4.2, and indicates a slow (exponential) decay in the vibrational amplitude with time. Therefore, the vibration is not a pure sine wave, but a decreasing sine wave. If the frequency content of this waveform is analysed, we find that in addition to the pure natural frequency $\omega_0$ there are other, nearby frequencies mixed in. The frequency distribution in fact resembles Fig. 4.1. The greater the damping forces, the shorter the damping time $2\tau$ required for the vibrational amplitude to fall to $e^{-1} \approx 0.37$ of its initial value. Thus, in a highly damped oscillator, the distortion from a pure sine wave is greater, and the admixture of other

Fig. 4.2. Damped oscillation. The vibration amplitude decays exponentially, with a 'half-life' $2\tau$, depending on the strength of the damping forces. The frequency content of this decaying waveform has a spectrum similar to Fig. 4.1, with a spread $\Delta\omega \approx \tau^{-1}$.

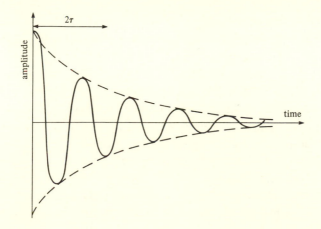

frequencies extends over a wider range. Theory shows that the frequency spread is approximately $\tau^{-1}$, so large $\tau$ (low damping) implies a sharply peaked frequency distribution.

These considerations also apply to the reverse process, i.e., response of the oscillator to a driving force. Thus, with larger damping (low $\tau$), the oscillator responds vigorously to a wider range ($\sim\tau^{-1}$) of frequencies.

If we assume very low damping, and hence a sharply peaked $\sigma(\omega)$, then the incoming flux from the thermal radiation will not vary appreciably in the narrow range $\Delta\omega \approx \tau^{-1}$, so we may remove the term 'flux per Hz/$2\pi$' from the integral in (4.2) and replace it by its value at $\omega = \omega_0$. Thus, dividing out this quantity we obtain

$$\text{integrated cross-section} \equiv \int \sigma \, d\omega$$

$$= \frac{(\text{outgoing rate of radiation})}{(\text{incoming flux per Hz}/2\pi \text{ at } \omega = \omega_0)}. \qquad (4.3)$$

We see that the cross-section, which can be envisaged physically as the effective area presented by the detector to the waves, is simply the ratio of power radiated to power intersected per unit area.

The power radiated is given by expression (3.1) with $\alpha = 4/15$. Let us write (3.1) as

$$\text{power radiated} = \left(\frac{4G}{15c^5}\right) M \, L^2 \, \omega_0^4 \times (\text{energy of vibration}),$$

where the vibrational energy of the oscillator, $Mx_0^2\omega_0^2$, is factored out. (The factor $Mx_0^2\omega_0^2$ is twice the average kinetic energy of the oscillator, i.e., $2 \times (\frac{1}{2} \cdot 2M \cdot \frac{1}{4}\dot{x}^2)_{\text{average}}$, where $\dot{x} = x_0\omega_0 \cos \omega_0 t$ and we use $(\cos^2 \omega_0 t)_{\text{average}} = \frac{1}{2}$. The overall factor of 2 comes from the fact that the total energy is made up of equal average quantities of kinetic and potential energy.)

From the meaning of thermodynamic equilibrium we know that the total energy is shared equally among all the available motions of the system (i.e., equipartition of energy between the various degrees of freedom). That is, the detector oscillations have the same average energy as each of the gravitational field oscillators (this energy being equal to $kT$, where $k$ is Boltzmann's constant). We may therefore write (4.3) in the following way:

$$\int \sigma \, d\omega = \frac{\left(\dfrac{4G}{15c^5}\right) M \, L^2 \, \omega_0{}^4 \times (\text{energy of vibration})}{\left(\begin{array}{c}\text{number of gravitational field} \\ \text{oscillators per unit} \\ \text{area per second per Hz}/2\pi\end{array}\right) \times (\text{energy of vibration})}$$

$$= \frac{4G \, M \, L^2 \, \omega_0{}^4/15c^5}{\left(\begin{array}{c}\text{number of wave oscillators} \\ \text{per unit area per second per Hz}/2\pi\end{array}\right)}. \tag{4.4}$$

The only remaining task is to calculate the number of wave oscillators in the denominator of (4.4). This task is straightforward, but the calculation is relegated to the Appendix. The answer is $\omega_0{}^2/\pi^2 c^2$, at frequency $\omega_0$. Inserting this factor in (4.4) yields

$$\int \sigma \, d\omega = \frac{4\pi^2}{15} \left(\frac{G}{c^3}\right) M L^2 \omega_0{}^2. \tag{4.5}$$

So long as $\sigma$ is appreciable only in the narrow range $\omega_0 \pm \frac{1}{2}\tau^{-1}$, the left-hand side (which is the area under the curve in Fig. 4.1) can be approximated by $\sigma_0 \, \tau^{-1}$, where $\sigma_0$ is the peak cross-section, at resonance ($\sigma_0 \equiv \sigma(\omega_0)$). A more careful analysis of the shape of Fig. 4.1 (see Appendix) gives $(\pi/2)\sigma_0 \, \tau^{-1}$. Thus, for a weakly damped oscillator

$$\sigma_0 \approx \frac{8\pi}{15} \left(\frac{G}{c^3}\right) M L^2 \omega_0{}^2 \tau. \tag{4.6}$$

Although equations (4.5) and (4.6) have been derived for the special case of the idealized detector shown in Fig. 2.14, the results are quite general, and in particular apply also to a solid block of metal. The dimensionless factor $\omega_0 \, \tau$ is known as the *quality factor*, or $Q$, of the oscillator, and is a measure of how finely tuned the system is. Alternatively, it is a measure of the amount of damping per cycle of vibration. A high $Q$ oscillator requires many cycles before it is appreciably damped.

It is interesting to express the cross-section in terms of $Q$ and the *gravitational* radius of the detector ($r_g = 4GM/c^2$ for a total mass $2M$):

$$\sigma_0 = \frac{4\pi^2}{15} \left(\frac{r_g}{\lambda_0}\right) Q \, L^2,$$

where $\lambda_0$ is the gravitational wave wavelength at the resonant frequency $\omega_0$. But $L^2$ is roughly the geometrical cross-section (i.e., area) of the detector (differences in geometry only alter the expression by small numerical factors). Thus

$$\frac{\text{(gravity wave cross-section)}}{\text{(geometrical cross-section)}} \sim \frac{r_g}{\lambda_0} Q. \tag{4.7}$$

One can see immediately how inefficient gravity wave detectors must be, for the gravitational radius $r_g$ is incredibly minute for laboratory apparatus. For example, a block of aluminium weighing one tonne has $r_g \approx 10^{-24}$ m. On the other hand $\lambda_0$ might be typically 100 km (see section 3.6), so $r_g/\lambda_0 \approx 10^{-29}$. Even with a $Q$ of $10^6$ (which corresponds to as much as $10^6/2\pi$ cycles of oscillation before damping becomes appreciable) the gravity wave 'stopping power' of this cubic metre of aluminium is only about

$$10^{-23} \times \text{geometrical area} \approx 10^{-23} \times (1 \text{ m})^2$$
$$= 10^{-23} \text{ m}^2.$$

(Current detectors are still more than an order of magnitude below this.) Thus, the effective area that a cubic metre of *resonating* aluminium presents to a flux of gravity waves with frequency about $c/(100\text{km}) \sim 3000$ Hz is much less than the size of an atom!

To see what sort of vibration this tiny effect induces, let us suppose there is an incident flux of gravity waves of $100 \text{ J m}^{-2} \text{s}^{-1}$ concentrated in the narrow frequency range $1/(2\pi\tau) \approx 0.003$ Hz around the resonant frequency. This is the sort of flux one might expect from the collision of two black holes at the centre of our Galaxy, but such a narrow frequency range could only be expected from a periodic system such as a vibrating neutron star. With an effective stopping cross-section of only $10^{-23}$ m$^2$, the aluminium bar only takes up $10^{-21} \text{ J s}^{-1}$. Suppose this incident flux drives the aluminium bar continuously. The energy in the resonant vibrations builds up and up until an equilibrium is reached, when the rate of dissipation of energy as heat is exactly balanced by this uptake. At 3000 Hz, a bar with a $Q$ of $10^6$ would have a damping time of around 53s. In this duration the uptake of energy is about $5 \times 10^{-20}$ J, which will be the same as the rate of energy damping to heat. Thus, the bar must have a vibrational energy of about twice this value (as about half the energy is damped out each 53 s) or $10^{-19}$ J.

It is helpful to convert this minute vibrational energy into a value for the actual movement experienced by the bar. The relation between amplitude $x_0$, energy $E$ and total mass $M$ for a harmonic oscillator is

$$\text{energy} = \tfrac{1}{2} M \omega^2 x_0{}^2, \tag{4.8}$$

so

$$x_0 \approx \left[ \frac{2 \times 10^{-19}}{1 \text{ tonne} \times (2\pi \times 3000 \text{ Hz})^2} \right]^{\frac{1}{2}} \approx 10^{-15} \text{m}$$

which is about the diameter of one atomic nucleus, representing a fractional change in bar length of only one part per million billion.

As another example, suppose we consider the entire Earth as a detector. How much is our planet shaken by gravity waves? The Earth has a quadrupole vibrational period of 54 minutes, corresponding to gravitational waves with wavelength of a billion kilometres. The gravitational radius is 4 mm, so $\sigma \sim 10^{-15} Q \times$ (area of Earth) $\sim Q \text{m}^2$. For $Q \approx 400$, and a steady incoming flux at the resonant frequency of $1 \text{Jm}^{-2} \text{s}^{-1}$ (very optimistic), the Earth's vibrational energy is about $10^6$ J, which produces a displacement at the Earth's surface of roughly $10^{-7}$ m. These sorts of figures are minute by any standards.

Of course, a realistic source of waves is unlikely to be continuous and at a single frequency. Most of the more intense radiation is expected to take the form of sudden pulses from gravitational catastrophes, such as the collapse of a star core. These will have a broad range of frequencies. It may then be advantageous to build a low $Q$ detector which will respond to this broad band. The cross-section will vary reasonably slowly as a function of frequency, as will the intensity of the waves. The energy will no longer be taken up at a steady rate by the detector, so interest centres more on the total energy delivered to the detector by the pulse, which is

$$\int \sigma(\omega) \times (\text{total incoming energy per m}^2 \text{ per Hz}/2\pi) \, d\omega.$$

## 4.2 Present technology

In the previous section it was explained how the principle of detection of gravity waves is very simple: one just looks for sympathetic vibrations in a piece of matter. The detector oscillator can be anything at all, and the simplest antenna is merely a block of metal. All that is necessary is the technology to measure changes in the length of the block very much less than $10^{-15}$ m.

The first laboratory gravity wave antenna was built by Professor Joseph Weber of the University of Maryland. Professor Weber has pioneered the subject of gravity wave astronomy for over two decades. His favoured detector, upon which several subsequent devices have been modelled, was a cylindrical bar of aluminium, 1.53 m long, 0.66 m in diameter and weighing 1.4 tonnes. It had a natural frequency in the lowest mode of 1660 Hz.

Two crucial problems face any would-be gravity wave astronomer using a Weber-type bar. The first is to measure bar displacements of unprecedented minuteness. The second is to isolate the bar from other sources of disturbance that would normally induce vibrations many orders of magnitude greater than those due to gravity waves. For example, seismic waves would swamp any gravity wave bursts.

Some idea of the delicacy of the gravity wave vibrations can be obtained by comparison of the energy involved with that which would be absorbed by the bar from *electromagnetic* waves. Using the extremely optimistic estimate of $10^{-21}$ J s$^{-1}$ for uptake of energy, this is equivalent to the light energy that would be absorbed from a lighted candle held one hundred million kilometres away.

Weber tackled the noise problem by suspending the bar delicately on a fine wire in a vacuum, and standing the apparatus on acoustic filters. In addition he built two identical bars and located them several hundred kilometres apart (one at the University of Maryland, the other at the Argonne National Laboratory near Chicago). He then monitored the bars for coincident displacements. By this technique he could eliminate many random extraneous disturbances, which would be unlikely to afflict both devices simultaneously.

To measure the displacements of the bar, Weber used piezoelectric strain transducers fixed around the middle of the cylinder. These convert tiny movements of the bar into electric impulses that can be amplified and recorded. The electronic circuits were tuned to the bar's fundamental vibrational mode frequency of 1660 Hz, for maximum sensitivity.

A major obstacle to such high-sensitivity measurements is extraneous noise, not from the laboratory surroundings, but from inside the bar itself, caused by the bar's own atoms cavorting about. This internal noise is due to thermal agitation of the bar molecules, and can only be reduced by lowering the temperature. Fundamental thermodynamics predicts that the bar will acquire a vibrational energy from these inevitable internal motions (sometimes called Brownian motion) equal to, on average, $kT$, where $k$ is Boltzmann's constant and $T$ the bar temperature. At room temperature $kT$ is about $4 \times 10^{-21}$ J, which is comparable to the vibrational energy induced by the most powerful gravity waves that one can reasonably expect. The motion of the bar due to this noise is obtained from equation (4.8):

$$x_{\text{noise}} = \left( \frac{2kT}{M\omega^2} \right)^{\frac{1}{2}} \tag{4.9}$$
$$\approx 2 \times 10^{-16} \, \text{m}$$

in this case. It is a remarkable thought that even the minute sound of the bar atoms rattling around is likely to drown out the ringing caused by the ephemeral gravity waves.

Although the amplitude of the thermal motion is great, the characteristics of the motion may well differ from that induced by gravity waves. In particular, the thermal motion is not a periodic vibration, but a series of fluctuations. The rate of change of these fluctuations is slow compared with that produced by a short, sharp burst of gravity waves, such as one might expect from gravitational collapse, for example. This is because the thermal noise is due to the collective participation of vast numbers of atoms, and the likelihood of a large scale cooperation between many atoms to produce a sudden 'kick' is very low. Instead, the bar length wanders about at random, changing appreciably only on a time scale determined by the *damping time* of the bar. (There is an intimate connection between the damping properties of the bar, which convert the organized vibrational motion into heat, i.e., chaotic molecular motions, and the inverse process of random fluctuations in molecular chaos conspiring to produce large scale, organized motion.) On the short time scale $\Delta t$ of the pulse, the likely change in length due to noise is lower than (4.9) by a factor $(\Delta t/\tau)^{\frac{1}{2}}$, where $\tau$ is the damping time. Typical burst durations are between $10^{-3}$ s and $10^{-1}$ s, while $\tau$ might be in the region of 20 s, giving an *effective* $x_{\text{noise}}$ of around $10^{-17}$ m. For comparison, the bar displacement due to the collapse of a ten solar-mass star to a black hole at the centre of our Galaxy also works out at around $10^{-17}$ m.

Clearly, thermal noise can be reduced by cooling the antenna and increasing the damping time, i.e., the quality factor $Q$. Experiments are being planned in which the antenna is to be cooled to about $3 \times 10^{-3}$ K, giving an improvement in sensitivity by a factor of about 300. Moreover, the careful use of materials such as niobium enables $Q$ values of $10^7$–$10^9$ to be obtained. Better still, single crystals of sapphire, grown to a mass of many kilograms, could mean that $Q$s as high as $10^{14}$ might not be impossible, corresponding to damping times of $10^{13}$ oscillations, or about a *year* even at MHz frequencies!

In recent years great progress has been made in reducing thermal noise in the antenna towards the technically feasible goal of $10^{-21}$ m, which would be necessary to detect the much more frequent bursts of gravity waves expected from a large group of distant galaxies (such as the so-called Virgo cluster), rather than the once-per-decade black hole burst expected from within the Milky Way.

However, other noise problems afflict these delicate measurements because the sensors themselves are subject to thermal fluctuations, which, when amplified by the electronics, can drown the tiny signal. Sensor noise energy is proportional to the frequency bandwidth, so if the measurement time is made very long, this bandwidth can be made very narrow. On the other hand, if the sensor is to measure the vibrations of the bar, it must be coupled to the bar in some way, so the thermal motions of the sensor will react back on the bar, producing a force just like that of the antenna's own thermal motions. The longer the measurement takes, the greater this disturbance will be. There is thus a conflict between direct sensor noise, and the secondary effect that this noise induces by disturbing the antenna. The best compromise is to make both these disturbances about equal, so long as the measurement time still exceeds the burst duration.

Very sensitive detectors have been planned by Vladimir Braginsky of Moscow University. They consist of a microwave cavity – a cavity inside a conducting surface in which a standing wave pattern of electromagnetic microwaves is set up. One wall of the cavity consists of the antenna surface. As it jiggles about, so the electromagnetic wave pattern is disturbed.

One can ask whether there is any limit to the sensitivity of a gravity wave detector, or whether with sufficient time and money, any gravity waves, however feeble, could be detected. We have seen how the measurements to be made are so delicate that even atomic effects must be taken into account. When atomic limitations are approached, ordinary mechanics and electronics cease to apply. Instead, one must resort to the quantum theory. In chapter 3 it was explained how quantum fluctuations and noise produced gravity waves at the end of the Planck era. Quantum fluctuations will also occur, but on a much longer time scale, inside gravity wave detectors.

First let us investigate the quantum nature of the gravity wave itself. By analogy with the concept of photons, one may envisage the gravity wave as a shower of gravitons. At an angular frequency $\omega$, each graviton has an energy $\hbar\omega$. A supernova explosion in the Virgo cluster of galaxies (which, by cosmological standards, is rather close at about $10^7\,\mathrm{pc}$) would produce a total burst of energy at Earth of about $5 \times 10^{-6}\,\mathrm{J\,m^{-2}\,Hz^{-1}}$ over a bandwidth from about 1 Hz up to $10^3\,\mathrm{Hz}$. This corresponds to a total of about $10^{28}$ gravitons intercepting a Weber-type antenna, in a burst duration of about 0.3 s, which implies that there are around $10^{20}$ gravitons inside the detector at any one time during the burst. Clearly this number is so large that we do not have to worry about the quantum properties of gravity as such, which is fortunate as

*a*

*b*

there is as yet no acceptable theory of quantum gravity (see section 5.4).

However, when we take into account the fact that the cross-section of the detector may only be $10^{-25} m^2$, and that only about one graviton in $10^3$ will lie in the narrow ($\sim 1 Hz$ or less) bandwidth required to excite the detector, we see that only *one* graviton is likely to be absorbed, i.e., converted to a phonon. As quantum theory demands that the detector cannot absorb a fraction of a graviton, there is clearly a chance that the gravity wave will have *no effect at all* on the bar, i.e., no graviton will be absorbed. We have therefore reached a fundamental limit to the sensitivity of the detector. Sources of gravity waves less energetic than this will apparently never be detectable without substantially increasing the cross-section of the detector. As detector sensitivity must reach at least the level of the Virgo cluster if several events per year are to be expected, this quantum limit would seem to deal a death blow to gravity wave astronomy.

Although the quantum nature of the gravity wave can be ignored, it is obvious that the quantum nature of the antenna and sensor equipment must be considered. Braginsky has studied this point to see if the disturbing limit mentioned above can be circumvented. First it must be appreciated that quantum theory predicts the existence of discrete energy levels. These levels are normally unobservable in macroscopic objects, and only at atomic dimensions do they become manifest. Now we must consider these atomic-level effects in a tonne of metal! In a Weber-type bar, the energy levels are only $10^{-31} J$ apart (far less than in an atom). Therefore, the energy of one quantum ($\hbar\omega$) of bar excitation is almost inconceivably small – about the same as the energy that would be acquired by a single electron dropped from rest from a height of one centimetre (though fortunately most electronic impacts do

Fig. 4.3. (*a*) Professor G. Papini and the cryogenic detector being developed by his group at the University of Regina. The detector itself is a large, exceedingly pure quartz crystal, cooled to near absolute zero to reduce the effects of thermal noise. (*b*) The laboratory is situated below ground in a quiet part of the Canadian prairies. (*Photographs reproduced by courtesy of Professor G. Papini, University of Regina*) not deliver much energy to the bar at the low frequencies of interest.)

Using equation (4.8) we may calculate the vibrational displacement caused by a single quantum of excitation. Putting energy $= \hbar\omega$, we obtain

$$x = \left(\frac{2\hbar}{M\omega}\right)^{\frac{1}{2}}$$ (4.10)

$$\approx 3 \times 10^{-21} \, \text{m}$$

for a Weber-type bar. This is some three or four orders of magnitude beyond current sensor technology, but may not be impossible to achieve in the foreseeable future. It is also well below the noise level of (4.9), so powerful refrigeration techniques will be necessary if gravity wave telescopes are even to reach part of the way to the Virgo cluster.

The ultimate objective of sensor technology is to detect the length change in the bar caused by a single quantum jump. If the bar is initially undisturbed, this would enable the limit (4.10) to be achieved. However, Braginsky has pointed out one way of beating the limit of (4.10). Suppose that, instead of starting out undisturbed, the antenna is initially vibrating in a highly excited state. If this state contains $n$ quanta of excitation, it will produce a displacement of $(2n\hbar/M\omega)^{\frac{1}{2}}$, i.e., $\sqrt{n}$ times (4.10). Suppose the gravity wave now induces a jump of one quantum when the bar is in this excited condition. The new value of $x$ will then be $[2(n+1)\hbar/M\omega]^{\frac{1}{2}}$. Consequently the *change* in $x$ when this jump occurs is

$$\left(\frac{2\hbar}{M\omega}\right)^{\frac{1}{2}} [\sqrt{(n+1)} - \sqrt{n}] \approx \frac{1}{\sqrt{n}} \left(\frac{\hbar}{2M\omega}\right)^{\frac{1}{2}}$$ (4.11)

for large $n$. Thus, by this manoeuvre, the quantum limit (4.10) can be converted into the smaller limit (4.11) enabling smaller changes in bar length to the measured.

In order to exploit this improvement, it is necessary to detect the single quantum jump among the $n$ other quanta. This means measuring the number of quanta in the bar to exact precision. The feasibility of such a task requires an analysis of quantum measurement devices. It was shown by the Canadian physicist William Unruh that sensors using linear couplings can only measure $n$ quanta to within a factor $\sqrt{n}$ – thus wiping out the $1/\sqrt{n}$ improvement factor in (4.11). Fortunately it seems that more complicated couplings may enable (in principle at least) exact precision. However, in order to avoid conflict with the Heisenberg uncertainty principle, it is necessary to forsake all information about the phase of the vibrations.

## 4.3     Other detectors

Some attention has been devoted to the question of the detection of continuous gravitational radiation, rather than short pulses. A steady flux should emanate from pulsars and binary stars, as discussed in chapter 3. Because these incoming waves can be monitored for an extended period, more total energy can be pumped into the antenna. However, there are deep problems here too. A simple bar detector would need to be enormous to resonate at the low frequencies involved ($\sim 10\,\mathrm{Hz}$), so some more complicated design is necessary. A detector of this sort, tuned to the Crab pulsar, was built by a Japanese group to operate at $60.2\,\mathrm{Hz}$. Unfortunately, the limited sensitivity of their equipment only enables an upper limit to be placed on the quantity of gravitational radiation from this source; this limit is 200 000 times greater than that which can be deduced from other considerations.

Although the initial effort, and much of the subsequent work, was directed towards the detection of gravity waves using Weber-type suspended cylinders, other devices are either being planned or are under construction.

One of the most promising uses the principle of the interferometer. One possible arrangement is shown in Fig. 4.4 and consists of a laser and two mirrors, $M_1$ and $M_2$, carefully suspended. The coherent laser light is split into two, one beam bouncing back and forth many times between $M_1$ and the central mirror $C$, the other beam doing the same on the $M_2$ 'arm'. Eventually the two beams are recombined and an interference pattern results where the peaks and troughs of the respective light waves overlap. If a gravitational wave comes along, the mirrors will move, and the interference pattern will wobble.

An advantage of this system is that it is non-resonant (the natural frequencies of the suspended mirrors being much less than that of the gravitational waves). Consequently, instead of responding to just one frequency in the waves, the mirrors simply follow the wave pattern whatever its shape. Thus, one does not merely register the passage of the wave, but also obtains information about the wave characteristics.

The first such detector achieved a sensitivity of only about 1 per cent of the Weber-type bars, but by making the arms of the interferometer very large, this sensitivity can be improved. Also, by using high-quality reflective surfaces, up to 300 reflections could be achieved, implying an effective arm length of, say, $3 \times 10^5$ m. These detectors could prove to be superior to the resonant bars at low frequencies, though the problems with outside noise (e.g., seismic waves) below 0.1 Hz might necessitate the use of space-borne equipment.

Another proposal for a non-resonant detector is to use distant spacecraft. If a gravity wave in the frequency range $10-10^{-4}$ Hz (wavelengths from about one Earth-diameter up to the size of the Solar System, such as would be produced by a superhole in another

Fig. 4.4. Laser interferometer. Schematic diagram showing a laser beam split into two, with each sub-beam bouncing back and forth many times between mirrors before being recombined with the other to produce an interference pattern. Minute changes in the inter-mirror distances produce shifts of the pattern.

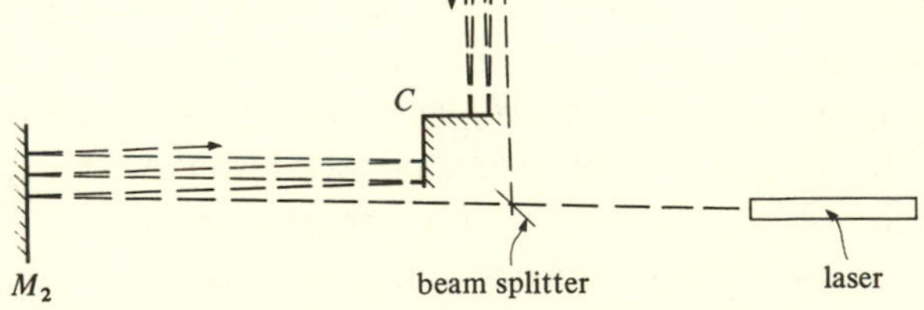

galaxy) passes through the Solar System, it will wobble the space-craft (and the Earth). This tiny motion could be detected using the technique employed by police on car speed radar traps. When a radio wave bounces off a receding object, its 'echo' has a slightly lower frequency owing to the fact that the reflecting surface is receding. This Doppler 'redshift' effect is routinely used to monitor the movements of spacecraft, and can achieve an accuracy in frequency measurement of about 3 parts in $10^{13}$, though this could be improved.

The procedure would be to use a highly frequency-stable maser, or a superconducting cavity, as a reference oscillator, and to beam the oscillations continuously to a spacecraft. The latter would be equipped with a transponder to send back the echo, and the outgoing and return signals would be compared. As the gravity wave pulse passes it briefly wobbles first the Earth, then the spacecraft (or vice versa). The Earth-wobble moves both the receiver, thereby temporarily Doppler-shifting the return beam, and the transmitter, which simultaneously shifts the outgoing beam. When the spacecraft is wobbled another temporary Doppler-shift is imparted to the return beam. Finally the Earth-wobble imprinted in the outgoing beam returns in the transponder echo, complete with wobble. Thus the gravity waveform is encoded three times in the signals, making it easier to spot among the noise. Moreover, by comparing the delay between the signals, some information about the direction of the source can be obtained.

A major source of noise is the instability of the reference oscillator; another is the effect of the tenuous interplanetary medium. Calculations indicate that gravity wave bursts from $10^7$ solar-mass superholes might be expected to induce Doppler frequency shifts of one part in $10^{16}$ or $10^{17}$ at around $10^{-3}$ Hz. This is about 1 per cent of the present noise level.

Also within reach of present technology is the possibility of detecting a continuous background of gravity waves, either from distant sources or from the big bang. If the same quantity of energy as is contained in the electromagnetic primeval heat background $(4 \times 10^{-14} \mathrm{J\,m^{-3}})$ were present among gravity waves with wave-lengths around 1 AU (such long wavelengths would be unlikely to originate in the conventional big bang scenario, but might have another origin), then they might be detected in the near future.

Since the original experiments of Weber around 1970, groups all over the world have followed his pioneering lead to develop their own gravity wave telescopes. The technological problems they have faced have been formidable. In this chapter, some of the difficulties encountered in the measurement of tiny disturbances have been outlined, but equally daunting problems afflict other aspects of the experiments – problems of good isolation, electronics and data handling, achieving low temperatures, building delicate suspension systems, machining accurate bars or growing pure crystals, and much else. Tremendous progress has been made, but much more is necessary if we are to go beyond the mere detection of gravity waves. The ultimate goal is to provide gravity telescopes with observational potential to rival, say, X-ray, or even radio telescopes. Improvements in sensitivity of several orders of magnitude will be necessary before this goal is achieved.

In this book, very little has been said about the development of the peripheral equipment, or the imaginative proposals for other designs of detectors, such as the graviton–photon resonant converter. Whatever the eventual achievements of gravity detectors themselves, there is no doubt that the expertise acquired from this sustained venture into ultra-high-sensitivity measurements will prove to be of enormous benefit to science and technology as a whole.

# 5 Have they been seen?

In the 16 June 1969 issue of *Physical Review Letters*, Professor Joseph Weber of the University of Maryland, the man whose vision and tenacity provided much of the early stimulus in the search for gravity waves, published a dramatic announcement. His two suspended-bar antennae, at Maryland and the Argonne National Laboratory, were recording coincident pulses of an unknown origin at the rate of about one per day. If correct, Professor Weber's results were one of the most exciting, yet astounding, scientific discoveries since the Second World War. More than anything else the 'Weber events', as they came to be known, sparked off the race for ever-better gravity wave detectors.

## 5.1 The Weber events

In the late 1960s, Joseph Weber began to record vibrations on a simple resonant bar detector, of the type described in the previous chapter, stationed at College Park, Maryland. The tremors, which had about three times the thermal energy of the bar, apparently occurred on a few dozen occasions each day. Mindful of the many sources of disturbance more prosaic than gravity waves, he set up a parallel experiment at the Argonne National Laboratory near Chicago, and encountered a similar rate of disturbance there. However, only about one event per day seemed to occur simultaneously on both bars, but from a statistical analysis Weber concluded that there was little likelihood of these all being chance conjunctions.

Of course, given two sets of random pulses, *some* coincidences are bound to occur by chance, but Weber estimated the proportion of chance coincidences to be an order of magnitude less. One check on this estimate was to introduce a time delay into one of the electronic circuits that monitored the pulses, to see if the rate of coincident pulses fell. Apparently it did. As events caused by gravity waves would occur simultaneously at the two locations to within the time resolution of the equipment ($> 0.1\,\mathrm{s}$), this procedure would automatically eliminate all gravity wave induced disturbances.

Weber made a careful study of all the influences other than gravity waves that could have been responsible for the events. These included solar flares, lightning, artificial radio signals and electric activity, and seismic waves. None of these seemed to be responsible. Nor could sudden 'quakes' due to temperature changes in the bars be the explanation, for they would be most unlikely to occur simultaneously in both.

Perhaps the most intriguing announcement of all was the so-called 'sidereal anisotropy'. Although the bars remain fixed in the laboratory, as the Earth rotates the cylinder axes turn relative to the distant stars. Because of their cylindrical shape, these bars are most sensitive to gravity waves when the direction of propagation of the waves is perpendicular to the axes. If the waves are coming from a fixed direction in space, one would expect a variation in intensity of the bar vibrations over a one-day period.

There is subtlety, however. Not only does the Earth turn on its axis; it also revolves in its orbit around the Sun. In one day, this additional orbital rotation amounts to about four minutes' worth of axial spin. Thus, the rotation period for the Earth relative to the distant stars is about four minutes short of that relative to the Sun. (In one year all these four-minute differences add up to a whole day.) The former period is called the sidereal day, while the latter (24 hours) is the solar day. Their difference is manifested by the slow daily migration of the Sun's position against the background constellations, caused by our gradually changing line of sight as the Earth's orbital motion alters our aspect.

Any man-made activity might be expected to show a 24-hour (solar day) periodicity, for the only people who keep sidereal day habits are astronomers. On the other hand, any activity originating in galactic space or beyond would have a sidereal variation, as its location would be fixed relative to the distant stars, not the Sun. Weber divided his detector events into three sets representing four-hour intervals, and found that in about 150 events recorded over a six-month period there was a significant correlation with sidereal time, but not solar time. The events seemed to be originating in

Fig. 5.1. Weber's gravity wave detector. Joseph Weber is shown attending to one of the aluminium cylinders used in his early work on gravity wave detection. Bars of this type are delicately suspended in a loop of wire, and the strain transducers round the middle record any vibrations. A decade ago Weber claimed to detect coincident tremors in two such devices several hundred kilometres apart. (*Photograph reproduced by courtesy of Professor J. Weber, University of Maryland*)

deep space. Curiously, the rise and fall of the bursts showed a twice-daily periodicity. This is expected because one maximum, when the bars are perpendicular to the waves with the source on one side of the Earth, should be followed about twelve hours later by another maximum when the bars are again perpendicular, but with the waves coming from the other side of the Earth (the Earth itself is virtually transparent to the waves). By observing the times of the maxima, Weber concluded that the waves were emanating from the centre of the Milky Way.

These observations by Weber were greeted with great excitement and interest in the early 1970s, but also with some scepticism. Superficially at least, the experimental results seemed very compelling, and the sidereal anisotropy was exactly the sort of effect that one would expect if the Weber events were being caused by violent black hole encounters in the crowded centre of our Galaxy. However, Weber seemed to be a victim of his own success, for if *all* his reported events were due to gravity waves, then the level of activity was very much in excess of what was expected.

In chapter 3 it was explained how the collapse to a black hole of a ten-solar-mass star at the centre of our Galaxy would produce a flux at Earth of around $10^5 \, \mathrm{J\,m^{-2}}$, which would induce a vibration in a one metre bar of only about $10^{-17} \, \mathrm{m}$. It is not clear if even this energetic an event would have been detectable with Weber's original equipment, but, even if it was, most astronomers estimate that such an occurrence would have a frequency of perhaps two or three per *century* (the frequency of supernovae per galaxy), whereas Weber's events were more like one per *day*. Moreover, as the bandwidth of his detectors was only about 0.1 Hz, even this high rate would presumably only represent a tiny fraction of the total number of bursts at all frequencies.

Of course, one can envisage many other sources of gravity wave pulses that might be as frequent as this; for example, a superhole at the centre of the Galaxy with a mass of about $10^7$ solar masses, this superhole swallowing a new star about once a day, with each devoured star liberating a good fraction of its rest mass as gravity waves. However, if this rate of mass loss were even remotely typical, the Galaxy would be losing mass into intergalactic space in the form of gravitational radiation at a prodigious rate, and it is difficult to see how the effects would not show up on the structure of the Galaxy over billions of years.

Whatever the source of Weber's events, it became imperative to verify them in other laboratories. In the early 1970s several research groups in various countries set up similar experiments using bars based on Weber's pioneering design. Some of these groups claimed a significant improvement in sensitivity over Weber's original apparatus. Apart from a single impressive event reported by Professor R. Drever of Glasgow University in 1972, none of these groups obtained results that could be interpreted as strong evidence for gravity waves.

For a while controversy raged over the status of Weber's events, the technical aspects of his and other groups' equipment, and the analysis of the data. Some of the debates were conducted with perhaps less courtesy than is usual in scientific affairs. Today, almost a decade later, there seems to be general agreement that a considerable improvement in sensitivity is still necessary if routine astronomical events are to be monitored using gravity waves.

The Weber events themselves have never been adequately explained. While some people evidently believe that they are purely spurious – perhaps a product of the computer – it has to be conceded that Weber may simply have been fortunate enough to go 'on the air' at a time of unusually intense gravity wave activity. Perhaps there is a superhole at the centre of the Galaxy and from time to time clusters of stars drop into it. If these events are extremely rare, and Weber just happened to spot one such set, then the mass loss problem would not be so severe. It is hard to see how we can know the answer to these conjectures until a great deal more is known, both about gravity waves and the conditions at the centre of our Galaxy. Until then, the Weber events must stand as an important and intriguing milestone in the development of gravity wave astronomy.

## 5.2    The back reaction

Although direct detection of gravity waves by a laboratory 'telescope' is the most satisfactory procedure for exploiting the information content of gravitational radiation, it is not the only way in which the existence and strength of this radiation can be inferred. In chapter 3, the power output of a variety of sources was discussed, and in some cases it was found that a significant fraction of the total available mass–energy could be radiated away. The loss of all this energy by the source cannot occur without some drastic effects on the source itself. Even if we cannot yet spot the gravity waves coming off, we might still be able to see the reaction on the source.

Gravity wave emission is expected to have a strong damping effect on, for example, the violent oscillations in shape that would accompany the birth of a neutron star. Less dramatic, but easier to study, are the effects in a binary star system. Here, as the two stars orbit around their common centre of gravity, the energy used to produce the gravitational radiation must be supplied by the orbital motion. The steady power drain will therefore cause the stars to spiral slowly together.

To estimate the rate at which the orbit decays, return to equation (3.10) giving the output from a binary star system with two equal mass stars in circular orbit. The characteristic time scale for an appreciable change in the orbital radius is

$$\tau \equiv \frac{\text{orbital kinetic energy}}{\text{power}} = \frac{GM^2/2R}{(64G^4/5c^5)(M/R)^5} = \frac{5}{128} \frac{c^5}{G^3} \frac{R^4}{M^3}. \tag{5.1}$$

In Table 3.1 on page 70, $\tau$ is given alongside a number of binary star systems. Most of the values run from billions of years upwards.

Note how sensitive the decay time is to differences in the orbital radius $R$. A decrease of a factor 10 in radius will produce a 10 000-fold precipitation in the collapse of the orbit. If (5.1) is re-expressed in terms of the orbital period $P$ (in seconds), so that

$$\tau \sim \frac{c^5 P^{8/3}}{(GM)^{5/3}}, \tag{5.2}$$

it can be seen that a solar-mass binary system with a period of only a few hours would decay in a time comparable to the age of the Universe ($10^{10}$ years). Contrast this with the enormous time ($10^{23}$ years) required for the Solar System, which radiates gravity waves at a power of a few kilowatts, to collapse as a result.

It is also amusing to calculate the lifetime of the lightest known binary system, i.e., positronium. Consisting of an electron and its antimatter counterpart, the positron, positronium is annihilated electromagnetically into two photons in about $10^{-10}$s. Quantum theory gives the energy of the lowest orbit as $me^4/4h^2$, where $e$ is the electronic charge and $m$ the mass of each particle. To obtain a characteristic time, we take $h/$(energy), so the power flow is roughly $m^2e^8/h^5$. Using equation (3.8), one finds for the gravity wave power output $Gm^4e^{16}/h^{10}c^5$, and a lifetime of energy/power $\sim h^8c^5/Gm^3e^{12}$, which is around $10^{30}$ years, or $10^{47}$ times longer than the corresponding electromagnetic process.

It was mentioned in chapter 3 that ordinary stars cannot approach too closely without disruption from tidal effects, but doubly-compact stars (i.e., a binary system in which both stars are collapsed) could orbit 'cleanly' within $10^6$ km. How common are doubly-compact binary stars? The problem is complicated because such systems are continually forming, while others are decaying and disappearing. When these stars get too close to each other, tidal processes eventually take over and the system becomes violently disrupted, probably ending up as a single object surrounded by debris, or as a black hole. Estimates range from one new system appearing every few thousand to every few hundred thousand years in our Galaxy. Knowing the decay rate, one can estimate the likelihood of finding such a system with a given period. It seems reasonable that somewhere in the Galaxy there is one doubly-compact binary with an orbital period of only a very few minutes. Such a system would decay noticeably in a few thousand years, and produce a gravity wave flux at Earth of perhaps $10^{-12} \mathrm{J} \mathrm{m}^{-2} \mathrm{s}^{-1}$. Going out as far as the Virgo cluster, one would expect there to be a system with a period of only a few dozen seconds, and a lifetime of only about a hundred years.

Although these results seem very exciting, a fundamental problem intervenes in any attempt to search for these systems. By their very nature, compact objects are hard to detect. A black hole, for example, is black, and only its gravitational effect on a nearby visible object, or perhaps X-rays produced by material falling into it, will give it away. A neutron star too is not visible directly.

Fortunately, there can be a very effective mechanism that enables a neutron star not only to be spotted, but its motion to be monitored in detail: the pulsar.

### 5.3    The binary pulsar

All stars rotate more or less rapidly: the Sun spins about once every 25 days. As a star shrinks, it must increase its rate of rotation to conserve angular momentum. During a supernova explosion, the core of the stricken star is likely to implode to only a few kilometres in radius, and the increase in rotation rate is spectacular. A typical neutron star might be born spinning dozens of times a second. The Crab Nebula, remnant of the 1054 supernova, contains such a neutron star.

Although made mainly of neutrons, the star will still contain some free electric charges and have a powerful magnetic field locked into it, estimated at $10^{12}$ gauss at the surface (compare with one-half gauss for the Earth's magnetic field). As the star rotates, so the magnetic field lines are whirled around, rather like the spokes of a wheel. In the midst of all this, electrons are stripped from the star's surface by a radial electric field, and spew out into the surrounding magnetic field like a sort of wind. As they tangle with the rotating magnetic field lines, the electrons are dragged around, their speed approaching that of light on the periphery of the 'wheel'. Travelling in circular paths, these electrons undergo enormous accelerations, which cause them to emit electromagnetic radiation, such as light and radio waves. Known as synchrotron radiation (because it is a phenomenon which also occurs in subatomic particle accelerators called synchrotrons), these waves are concentrated into a small cone of angles by the high velocities involved, so they emit a narrow beam of radiation rather like a searchlight. As the system rotates, this beam sweeps around the Galaxy, so that every time it passes Earth we see a pulse of radiation.

In 1967, the Cambridge astronomer Anthony Hewish and his student Jocelyn Bell (now Burnell) discovered the first of these pulsating radio sources, or pulsars. Soon, dozens more were found, including one in the middle of the Crab Nebula. The radio pulses

Fig. 5.2. Arecibo radio telescope, Puerto Rico. A segment of the dish, which is built into a natural crater, is shown. Radio waves reflected from its surface are focussed towards the suspended triangular framework that supports the detector. This gigantic instrument has been used to monitor the radio signals from the binary pulsar PSR 1913 + 16, whose pulses apparently reveal the emission of gravity waves from the system. (*Cornell University, Arecibo Ionospheric Observatory, Puerto Rico*)

are exceedingly regular, and can occur many times a second. By timing the pulses carefully, the central neutron star can be detected gradually slowing down in its rotation rate as it loses energy.

During 1974, two astronomers, Russell Hulse and Joseph Taylor of the University of Massachusetts, were using the giant radio telescope at the Arecibo Observatory of Puerto Rico to search systematically for pulsars. Many new ones were found, but one in particular caught their attention. Denoted PSR 1913+16, it is a very rapid pulsar, with a period of only 59 ms. What was peculiar about this object was not so much the rapidity of its pulses as the fact that the period seemed to be drifting noticeably. The equipment enabled the pulse period to be measured to within a microsecond, but PSR 1913+16 was changing its period by up to 80 $\mu$s a day, and even as much as 8 $\mu$s in five minutes.

The inevitable conclusion was that the pulsar is changing its motion rapidly, and that the variation in pulse rate is caused by the changing Doppler effect. The pattern of change fits precisely the model of a doubly-compact binary system, with the neutron star (pulsar) in orbit around another unobtrusive object, perhaps a black hole or another neutron star. As the pulsar revolves round its companion it alternately advances towards us and recedes. When advancing the radio waves receive a boost to their frequency (though not to their speed – recall the special theory of relativity!), and when receding the waves are drawn out at a lower frequency. These changes are reflected in the observed variations of the pulse rates received at Earth.

Not only had Hulse and Taylor discovered a closely-bound binary, but one with a built-in clock (the pulsar mechanism), which could be used to follow changes in motion and gravitational field to test the theory of relativity. The discovery caused considerable excitement, and the new pulsar was monitored carefully in the months that followed.

By combining their theoretical model calculations with the observations, and using the radio pulse arrival times to correct the model parameters, the astronomers were soon able to supply extremely detailed and accurate information about the PSR 1913+16 system. Some of the vital statistics are shown in Table 5.1.

Several features are worth noticing. First the *orbital* period of the binary is very short – a little under eight hours (an eight-hour 'year'). This means the orbital speed is high and the gravitational potential strong (both an order of magnitude greater than the corresponding values for the planet Mercury in the Solar System). This means that relativistic effects, which have so far only been measured in the Solar System where they are extremely small, can easily be measured. For example, the slight twisting of the ellipticity of the orbit (the advance of the periastron), which for Mercury amounts to a mere 43 seconds of arc per century, is a staggering four degrees a year for the binary pulsar. The fact that such relativistic effects are so large enables still more subtle, higher order relativistic 'corrections' to be measured. Therefore, the information content encoded in the pulses is very high, and it is this fact that brings the possibility of observing the effects of gravitational radiation within reach. Using the general theory of relativity to determine the orbital characteristics, it is possible to check the consistency of the model within this theory for the presence of gravity waves emanating from the system.

As discussed in the previous section, the rotation of one star around another should cause a gravity wave power drain leading to the slow decay of the orbit. For an eight-hour period, the decay time is less than $10^{10}$ years, leading to a change in the orbital period of a few parts in $10^9$ per year. With an inbuilt clock as accurate as a pulsar, changes of this magnitude would show up in a very few years.

Table 5.1. *Binary pulsar* PSR 1913 + 16

| | |
|---|---|
| Period of binary orbit | 27906·98172 s |
| Period of pulses | 0.059029995269 s |
| Rate of decrease of orbital period | $3.2 \times 10^{-12}$ s s$^{-1}$ |
| Rate of increase of pulse period | $8.64 \times 10^{-18}$ s s$^{-1}$ |
| Eccentricity of orbit | 0.617155 |
| Sky coordinates | 19 h 13 min 12.474 s, 16° 01′08″.02 |
| Advance of periastron | 4.226 deg yr$^{-1}$ |

Data taken from J. H. Taylor, L. A. Fowler and P. M. McCulloch, 'Measurements of general relativistic effects in the binary pulsar PSR1913 + 16', *Nature*, **277** (1979), 437–40.

At the Ninth Texas Symposium on Relativistic Astrophysics held in Munich in December 1978, Taylor announced what he claims to be the discovery of gravitational radiation. Fig. 5.3 summarizes his data, showing the observed orbital decay compared with the theoretical predictions using Einstein's general theory of relativity. The agreement seems very impressive. As time goes by and the orbit further decays, the comparison can be steadily improved in accuracy. It is hard to resist the impression that PSR 1913 + 16 really is emitting gravity waves.

Before this conclusion can be drawn, however, it has to be established that no other process is responsible for the orbital decay. One possibility is friction. If the binary system were polluted with gas and other debris, it would exert a drag on the orbiting stars and also cause orbital decay. Evidence against this is the absence of a varying frequency dispersion in the radio signals, which would inevitably occur if the radio pulses had occasionally to pass through substantial amounts of material.

Another cause of orbital decay is tidal friction. The neutron star's gravity will raise tides on the surface of its companion, and as the stars rotate about each other these tidal bulges will be dragged

Fig. 5.3. The points represent measured orbital phase errors caused by assuming a fixed value of the binary orbit period. Uncertainties associated with each point are comparable to or smaller than the point itself. The plotted curve corresponds to the orbital period derivative predicted by general relativity if the masses of the pulsar and its companion are equivalent and equal to 1.41 solar masses. (From J. H. Taylor, L. A. Fowler and P. M. McCulloch, *Nature*, **277** (1979), 437–40)

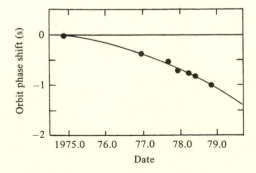

around. The effect of continually displacing vast quantities of viscous material is to convert the tidal energy into heat. The tidal bulges also exert their own gravity back on the neutron star, depleting its orbital energy by an amount equal to the heat generated by the tides. If the companion were an ordinary star like the Sun, this tidal decay would be colossal, so evidently the object must be much more compact. There is direct observational evidence to support this. The absence of any eclipse in the radio pulses, such as would occur if the pulsar disappeared behind its companion, limits the diameter of the latter to below $10^5$ km – too small to be an ordinary star. If it were another neutron star or a black hole, tidal effects would be negligible because of their tiny size. On the other hand, if the companion object were to be simply a shrunken star, such as a white dwarf or helium star (both are burnt out, low mass stars that could be larger than the Earth) then tidal effects could rival those due to gravity waves.

How can we determine the nature of the companion star? Some information is available from computing its mass, for if this were several solar masses, a black hole would be suggested (a gravitationally collapsed object with more than about three solar masses cannot, according to general relativity, avoid becoming a black hole). However, Taylor finds a mass of only about $1\frac{1}{2}$ solar masses, and favours a neutron star.

The estimated distance to the pulsar is 5000 pc, which is close enough for the presence of a helium star to be revealed by direct telescope search. In the summer of 1979, the results of such a search carried out at the Kitt Peak National Observatory were published in the journal *Nature*. The authors claim to have detected a star at the precise sky coordinates of the binary pulsar PSR 1913 + 16. The chances of this being an unconnected star, which just happens to lie in that direction, are estimated as 3 per cent. If subsequent spectroscopic work confirms that the object really is a helium star, it will considerably complicate our understanding of this binary system.

One final, but extremely important, point must be mentioned. In section 2.5 the reader was warned that the theoretical discussion of gravity waves would be based on an approximation that is frequently made in practice but which is open to question. Gravity is a non-linear theory: that is, gravity itself gravitates. This implies that, if two distributions of stress–energy–momentum produce certain gravitational fields, then the superposition of those two sources will not produce the superposition of the two fields, because the change in gravitational energy caused by the gravitational interaction between the two sources will add its own contribution to the total gravitational field. Gravity, in a sense, acts on itself, and in the foregoing chapters this non-linearity has been ignored.

Undoubtedly, this approximation is legitimate under many circumstances, so long as the gravitational fields of interest are weak enough, and in the subject of gravity wave detection this is probably true. When it comes to emission, however, we cannot be so sure. The field of the gravity waves close to an astronomical source is many orders of magnitude stronger than at Earth. Moreover, in a linear theory, a small change in the field can be assumed to produce a small reaction on the source. In a non-linear theory this cannot be assumed, and some physicists have claimed that the back reaction due to gravity wave emission could differ by large numerical factors from the naive estimates based on formula (3.3) and crude arguments about energy conservation. Taylor has argued that the good agreement between the binary pulsar data and the linearized theory based on (3.3) confirms the linear approximation to good accuracy. On the other hand, one could envisage that the agreement is purely coincidental, resulting from a conjunction of a much larger (or smaller) gravity wave back reaction combined with tidal effects on the companion.

Unfortunately, the non-linear theory of the back reaction under the emission of gravity waves has encountered severe problems of principle and technical complexity in the mathematics. Probably the discovery of the binary pulsar will stimulate a more energetic analysis of this longstanding theoretical problem.

### 5.4    Why gravity waves matter

The story of the search for gravity waves began with Einstein in 1918 changed from a theoretical to an experimental enterprise in the 1960s and moved towards a position of central interest to astronomers and physicists in the 1970s. Now, as we enter the 1980s, gravity wave physics is on the threshold of 'taking off' to become a major area of scientific enquiry. Given that gravity waves are probably far too weak ever to be used for telecommunications, why are they so important?

Some answers to this question have appeared among the text in the foregoing chapters. For the experimenter, gravity wave physics is exciting for two reasons. The first is the challenge of high precision technology. Being able to spot a movement of a million-billionth of a centimetre in a tonne of aluminium is an impressive achievement. In addition it is amazing to be approaching the limit of quantum engineering, where essentially subatomic effects are being manifested on the scale of the laboratory. The fact that physicists have had to take seriously the possibility that their sensors and probes might have to be so delicately planted on the antennae that they do not even disturb the equivalent of a single electron is a testimony to the way in which orders of magnitude have crumbled away under the ingenuity and expertise of modern experimenters. Whatever becomes of gravity wave physics, it would be surprising indeed if this level of precision did not 'spin off' into other areas of science.

To the astronomers, the experimental interest centres more on the subject of gravity wave astronomy – a new window on the Universe. In the last two or three decades, astronomers have opened up the electromagnetic spectrum from the visible and near infra-red and ultra-violet, first into radio waves, then X-rays, far infra-red and gamma rays. Each new wavelength range has brought a rich harvest of observations and increased our understanding of the cosmos enormously.

Electromagnetic waves all suffer, however, from a fundamental limitation: they cannot penetrate matter very far. Fortunately for astronomy, the Universe is rather empty, so much electromagnetic information propagates to us across billions of light years of space. Yet, because it cannot easily penetrate matter, it only carries information about the surface features of the source. Frequently the places of real interest to the physicist and astronomer lie buried deep inside matter, or are shrouded by gases or other material of less interest than the interior itself. Thus, when we look at the planets, or the Sun and stars, we only see their surfaces. The processes that power the stars lie concealed in their hearts. When we examine the microwave background heat radiation we are really looking, not at the big bang, but at the last surface of opaque gas about 100 000 years afterwards, when the cosmological material was still hot enough to be ionized.

The desire to probe into the inner vitals of the Universe, to examine matter and energy under the most extreme conditions, adds an additional interest to gravity wave astronomy, for this is not just a new region of the electromagnetic spectrum but a whole new spectrum. Like neutrinos, gravitons can easily pass through vast quantities of ultra-dense matter without being stopped, and can erupt out of the centres of stars, neutron stars and quasars, and from the very edges of black holes. Gravity waves can rumble across the Universe from the first conceivable moment of the big bang, and carry information about epochs that are more than fifty powers of ten earlier than the corresponding electromagnetic signals. Using gravity wave telescopes, astronomers could follow the tortured destruction of whole stars, the titanic explosions of quasars and the awesome collisions of black holes. They could 'see' the violent gyrations of nascent neutron stars, the turbulence of a supernova and the gentle but inexorable collapse of star clusters and galaxies. These details may still be a long way off, but it is hard to see how science can shrink from the new Maxwell–Hertz path laid before us by Einstein.

Turning to the theoretical aspects, even the mere discovery of gravity waves represents a landmark in the development of the theory of gravity. Einstein's general theory of relativity was published in 1915, but several decades later the theory is only beginning to be understood. Most physicists accept it as the correct description of spacetime structure and gravity, at least on a macroscopic scale. However, formidable mathematical and experimental problems are involved in verifying the theory. There are a number of reasons why this is so.

First, the extreme weakness of gravity means that few experiments can be done in the laboratory. Instead, the experimenter has to forsake control over the sources of gravity, and make use of what nature provides elsewhere in the Universe. At first sight it appears that the more massive systems offer the best promise. Unfortunately, the massive systems are also the most complicated. A galaxy, for example, with its $10^{11}$ stars, is unlikely to provide much usable information about general relativity. Instead, the Solar System has traditionally been the testing ground for the theory, and the so-called three classic tests of general relativity had all been successfully accomplished by 1919. Since then, advances in technology have largely been deployed in improving the accuracy of the three tests. Now, with the discovery of the binary pulsar, the first qualitatively new checks on general relativity since 1919 can be carried out.

Another reason for our slow progress in understanding general relativity is its mathematical complexity. Einstein's equations themselves are, in a formal sense, elegantly simple and concise. The problem comes when they have to be applied to the frequently complicated gravitating systems that are encountered in real astronomical situations. The non-linearity of the theory often spoils attempts to obtain exact solutions, while approximation methods can be long, tedious, and of dubious reliability. It has already been mentioned that the standard formulae for the emission of gravity waves and their reaction back on the emitter incorporate a linear approximation that has been questioned. With the experimental study of the binary pulsar, the validity of this crucial approximation can be tested.

Taking a wider perspective, gravity is one of four known fundamental forces of nature. The other three – electromagnetism and the weak and strong nuclear forces – all have quantum mechanical descriptions. Quantum theory began with the discovery that electromagnetic waves were emitted and absorbed in discrete quantities, or photons. If it is believed that the quantum theory must be applied consistently to all of physics (which is a virtually unanimous assumption) then gravity waves must also be quantized as *gravitons*.

The weakness of gravity means that experiments with gravitons, unlike those involving photons, are out of the question, so we can only check the theory that gravity is quantized by mathematical modelling: attempting to construct a comprehensive theory that will at least give sensible answers to physical problems, even though it may be beyond our technological capability to check them. Such a procedure cannot prove that quantum gravity is correct, but the absence of a viable theory might well call into question either quantum theory, general relativity, or both.

Attempts to formulate a viable theory of quantum gravity have been pursued with great energy for about twenty years, but without success. The problem is the following. According to the Heisenberg uncertainty principle of quantum theory, energy is not necessarily strictly conserved. Energy may be 'borrowed' for short durations so long as it is paid back promptly. The borrowed energy $\Delta E$ is related to the duration $\Delta t$ by

$$\Delta E \; \Delta t \sim \hbar.$$

The availability of these energy loans means that gravitons can be created out of nowhere, if they disappear again after a short enough time. Indeed, around every particle of matter is a whole beehive of these fleeting *virtual* gravitons, as they are called. The cloud of gravitons will interact with the particle and alter its mass–energy. On the face of it this correction ought to be very small, reflecting the weakness of quantum gravity effects. Instead, it turns out to be infinite. The reason for this is that there is no limit to how energetic the virtual gravitons might be, because if they are emitted and reabsorbed by the same particle, there is no lower limit to their lifetime $\Delta t$, so by the Heisenberg relation, $\Delta E$ can be arbitrarily large.

An infinite self-energy also occurs in the quantum theories of the other forces of nature, such as quantum electrodynamics. In the case of the latter, and in certain recent theories of the nuclear forces, these infinities do not spoil the predictive power of the theory because they do not occur in quantities that can actually be observed. Moreover, the number of infinite quantities is strictly limited.

Theories in which the infinities can be handled without ruining the sense of the theory are called *renormalizable*, and possess some very special mathematical properties to make them so. General relativity, unfortunately, does not seem to possess these properties, and is not renormalizable as it stands.

The failure to construct a straightforward theory of quantum gravity has led to some radical new approaches. One of these, called supergravity, seeks to combine gravitons with other types of quantum particles having half-integral spin (see page 13) in the hope that the *combined* infinities might be more amenable to renormalization. It is too soon to know if this approach will succeed.

Others have suggested abandoning general relativity altogether at the microscopic scale, while still more ambitious programmes attempt to build both spacetime structure and the quantum principles out of some more fundamental mathematical structure.

Whatever the outcome, the synthesis of gravity waves and quantum theory seems likely to have a profound impact on the future of physics.

# Appendix *Theory of gravity wave detectors*

When a sinusoidal gravity wave passes through a metal cylinder of the Weber type, it sets up sympathetic vibrations. The bar acts as a simple harmonic oscillator: when stretched, elastic forces produce a restoring force proportional to the stretching. Idealized, the system looks like Fig. 2.14(a). Thus, if the stretching is $x$, Newton's second law gives

$$\text{elastic force} = -Kx = 2M\ddot{x}, \tag{A.1}$$

where $K$ is a (positive) elastic constant and $M$ is the mass of each ball in Fig. 2.14(a).

Equation (A.1) may be solved

$$x = A\,\sin(\omega_0 t + \phi), \tag{A.2}$$

where $A$, $\phi$ are constants, and $\omega_0^2 = K/2M$. The frequency $\omega_0$ is the natural frequency of the bar, the frequency with which it will vibrate if struck suddenly and left free to oscillate.

As the bar moves, energy will be dissipated as heat. The bar therefore experiences a damping force proportional to the velocity $\dot{x}$. The damped oscillator obeys the equation

$$\begin{aligned} \text{total force} &= \text{elastic force} + \text{damping force} \\ &= -Kx - \gamma\dot{x} = 2M\ddot{x} \end{aligned} \tag{A.3}$$

($\gamma$ is a positive constant). This equation has the solution

$$x = A\,e^{-\gamma t/4M}\sin(\omega_1 t + \phi), \tag{A.4}$$

where $\omega_1 = (\omega_0^2 - \gamma^2/16M^2)^{\frac{1}{2}}$. This solution represents oscillations that decay in amplitude exponentially with a lifetime (i.e., e-folding time) $2\tau$ where $\tau = 2M/\gamma$, and in energy with a lifetime of $\tau$. Thus, the greater the damping constant $\gamma$, the shorter the life of the oscillations.

If, in addition to elastic and damping forces, the bar is also driven by a gravity wave, we must include a third force in equation (A.3). Ignoring angular factors, let us take a pure sinusoidal driving force $B \sin \omega t$, with $B$ constant. Then (A.3) is replaced by

$$\ddot{x} + \left(\frac{\gamma}{2M}\right)\dot{x} + \omega_0^2 x = \frac{B}{2M} \sin \omega t. \qquad (A.5)$$

This has the solution

$$x = \frac{(B/2M)\sin(\omega t - \theta)}{[(\omega^2 - \omega_0^2)^2 + \omega^2/\tau^2]^{\frac{1}{2}}} + A e^{-t/2\tau}\sin(\omega_1 t + \phi) \qquad (A.6)$$

where

$$\tan \theta = \frac{\omega/\tau}{(\omega_0^2 - \omega^2)}.$$

If the driving force continues for a time $\gg \tau$, the second, exponentially decaying, term can be neglected. The remaining term represents the oscillations of the bar driven at the imposed frequency $\omega$. If the incoming waves are close to the natural frequency, $\omega \approx \omega_0$, then $\omega^2 - \omega_0^2 \approx 2\omega_0(\omega - \omega_0)$ and (A.6) reduces to $(t \gg \tau)$ approximately

$$\left\{\frac{B/4M\omega_0}{[(\omega - \omega_0)^2 + (1/2\tau)^2]^{\frac{1}{2}}}\right\}\sin(\omega t - \theta). \qquad (A.7)$$

From (A.7) one sees that the amplitude of the sympathetic vibrations is a very sensitive function of $\omega$ in the vicinity $\omega = \omega_0$. Fig. 4.1 shows this functional dependence. The amplitude is a maximum at $\omega = \omega_0$, having the value $B\tau/2M\omega_0 = B/\gamma\omega_0$. The half-width of the peak is $\sqrt{3}/\tau$, so it is clear that, as the damping is reduced ($\tau$ increased), so the response function shown in Fig. 4.1 becomes narrower and the frequency selectivity of the bar increases. The rapid increase in response as $\omega$ approaches $\omega_0$ is called *resonance*.

The energy of vibration is equally shared on average between kinetic and elastic (potential) energy, so the total energy will be

$$E = 2M <\dot{x}^2>_{\text{average}} \approx \frac{B^2/16M}{(\omega - \omega_0)^2 + 1/4\tau^2}. \qquad (A.8)$$

This energy is dissipated inside the bar at a rate $E/\tau$. Because the vibrations described by (A.7) are steady, the bar must be absorbing energy from the gravity waves at rate $E/\tau$ also. This enables us to compute an expression for the *cross-section* $\sigma$ because the rate of energy absorption is $\sigma \times$ (incoming energy flux).

To relate the energy flux of the wave to the maximum force $B$ on the bar, note that the acceleration of the bar, $B/2M$, must be proportional to the length $L$, as it is a tidal effect (see page 46). The only other characteristic wave quantity is the frequency $\omega$, so on dimensional grounds

$$\frac{B}{M} \propto L\omega^2, \tag{A.9}$$

with the constant of proportionality depending on the strength of the wave.

As regards the energy flux, this too can only depend on $\omega$ and the physical constants $c$ and $G$. Once again on dimensional grounds we must have

$$\text{flux} \propto \frac{c^3\omega^2}{G}. \tag{A.10}$$

The energy flux must also depend on the strength of the wave, but, unlike the force $B$ which changes sign as the wave undulates, energy must always be positive, so the flux depends on the square of the wave amplitude and hence on $B^2$. Combining (A.9) and (A.10) with the correct dimensions yields

$$\text{flux} \propto \frac{c^3 B^2}{GM^2\omega^2 L^2}. \tag{A.11}$$

A full treatment using general relativity gives the constant of proportionality for unpolarized radiation as $15/32\pi$.

Dividing (A.8) by $\tau$ and equating with $\sigma \times$ (A.11) gives an expression for $\sigma$ near resonance:

$$\sigma = \frac{2\pi GM\omega_0^2 L^2/15c^3\tau}{(\omega - \omega_0)^2 + 1/4\tau^2}. \tag{A.12}$$

At resonance, $\omega = \omega_0$ and (A.12) reduces to equation (4.6), derived by a completely different method.

The integrated cross-section is

$$\int \sigma \; d\omega = \frac{2\pi GM\omega_0^2 L^2}{15c^3\tau} \times \int_{-\infty}^{\infty} \frac{d\omega}{(\omega - \omega_0)^2 + 1/4\tau^2} = \frac{4\pi^2 GM\omega_0^2 L^2}{15c^3}, \tag{A.13}$$

which is the same as equation (4.5).

·We end this discussion with a derivation of the formula used in section 4.1 for the number of gravitational field oscillators that produce waves crossing a unit area of the detector per unit time per Hz/$2\pi$.

First consider the simpler problem of waves on a stretched string. Suppose the end points of the string are fixed. Only certain wavelengths of vibration will now occur on the string, namely those for which an integral number of half-wavelengths just 'fits' onto the string (see Fig. A.1). Thus, the lowest mode of vibration has a wavelength $2L$, where $L$ is the length of the string, with corresponding angular frequency $\pi c/L$. The next-lowest mode has twice this frequency (half-wavelength), the next three times, and so on, corresponding to 2, 3, ... half-wavelengths on the string. For $n$ half-wavelengths, the frequency is $n(\pi c/L)$.

If $L$ is very large, the allowed frequencies are very closely spaced, so in the limit as $L \to \infty$ we may assume they are continuous and ask how many vibrational modes there are in the range $\omega$ to $\omega + d\omega$. Clearly, as $\omega = n\pi c/L$, the answer is $dn = (L/\pi c)d\omega$.

In three dimensions there are many more modes in the range $d\omega$,

Fig. A.1. The three lowest vibrational modes on a string of length $L$. The wavelengths are $2L$, $L$ and $2L/3$, and the frequencies $\pi c/L$, $2\pi c/L$ and $3\pi c/L$, respectively.

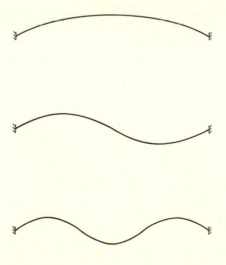

because vibrations can occur in three perpendicular directions. In a cube of elastic material of side $L$, there will no longer be one integer $n$ governing the mode frequencies, but three: $l$, $m$, $n$, corresponding to the three perpendicular directions. The allowed wavelengths are $2L/(l^2+m^2+n^2)^{\frac{1}{2}}$, and the allowed angular frequencies $\omega=(\pi c/L)(l^2+m^2+n^2)^{\frac{1}{2}}$, which is obtained by solving the wave equation subject to the boundary conditions that the wave amplitude vanishes at the surface of the cube.

The integers $l$, $m$, $n$ run over the set 0, 1, 2, 3, ... If they are plotted on a three-dimensional grid therefore, they all lie in the top right-hand octant. A given point $(l, m, n)$ in the grid lies at a distance $R\equiv(l^2+m^2+n^2)^{\frac{1}{2}}$ from the origin, so $\omega=\pi Rc/L$. If we consider a spherical shell lying between $R$ and $R+dR$ it will have a volume $4\pi R^2\,dR$, so in the octant of interest this is only $\frac{1}{8}\times4\pi R^2\,dR$. But $dR=L\,d\omega/\pi c$, so

$$\text{volume of shell}=\frac{\omega^2 L^3}{2\pi^2 c^3}\,d\omega.$$

The number density of points is one per unit volume of the grid, so the number of points in the shell is simply equal to the volume of the shell. Thus, the number of allowed vibrational frequencies in the range $\omega$ to $\omega+d\omega$ is $(\omega^2 L^3/2\pi^2 c^3)d\omega$. Because we are dealing with gravitational waves, which can have two independent polarization directions, we must double this quantity.

Now $L^3$ is the volume of the cube, so the number of wave vibrations per unit volume of space is $(\omega^2/\pi^2 c^3)d\omega$. The waves travel at speed $c$, which means that $c\times(\omega^2/\pi^2 c^3)d\omega$ waves cross each unit area of surface per second. Thus, the number of oscillators per unit area per second per Hz/$2\pi$ is $\omega^2/\pi^2 c^2$.

# Index